MODERN EARTH

# Study Guide

**HOLT, RINEHART AND WINSTON**

A Harcourt Classroom Education Company

**Austin** • New York • Orlando • Atlanta • San Francisco • Boston • Dallas • Toronto • London

# About the Modern Earth Science Study Guide

This Study Guide has been developed to help you succeed in your study of earth science using the textbook *Modern Earth Science.* The Study Guide focuses on the key concepts and terminology presented in each chapter of your textbook. The 30 chapters of the Study Guide correspond to the chapters in your textbook. As you work through each Chapter Review worksheet, you will encounter three types of exercises that relate to the learning objectives in the textbook. Each exercise is titled Multiple Choice, Critical Thinking, or Applications. Each type is briefly described below.

**Multiple Choice** The Multiple Choice questions test your understanding of both the concepts and the vocabulary introduced in each chapter. The Multiple Choice questions may ask you to interpret or identify a figure that relates to the chapter.

**Critical Thinking** The Critical Thinking questions allow you to synthesize information from the chapter while practicing your writing skills.

**Applications** The Applications questions ask you to use previously learned information or skills in a new context. The Application questions require putting bits of information together in new ways.

## How to Use This Study Guide

The Chapter Review worksheets are a valuable tool that can be used in a number of ways to guide you through your textbook. Depending on how your teacher plans your earth science course, the Chapter Review worksheets can be used as a prereading guide to each section, helping you to identify and review the main concepts of each chapter before and during your initial reading of each chapter. You can also use the worksheets after reading each chapter to test your understanding of the chapter's main concepts and terminology. Finally, you can use the Chapter Review worksheets to review for your earth science exams. Regardless of how you and your teacher use the *Modern Earth Science Study Guide,* it will help you to determine which topics you have learned well and which topics need further work.

## To the Teacher

Answers to the Study Guide Chapter Review Worksheets can be found in the Annotated Teacher's Edition of *Modern Earth Science.* The Chapter Commentary and Answers for each chapter contains an answer section for Chapter Review. The Study Guide Chapter Review Worksheets and Answers can also be found in the *Modern Earth Science* One-Stop Planner CD-ROM.



**Cover:** (earth and moon), Pictor/Uniphoto; (clouds and rainbow), Craig Aurness/Corbis ; (trees, water and bottom left fish), Dave Fleetham/Pacific Stock; (bottom right fish), Fred Bavendam/Peter Arnold, Inc.

Printed in the United States of America

ISBN 0-03-064314-7 1 2 3 4 5 6 7 129 05 04 03 02 01 00

# CONTENTS

# *Concept Mapping* *A Way to Bring Ideas Together*

## What Is a Concept Map?

Have you ever tried to tell someone about a book or a chapter you've just read, and you find that you can remember only a few isolated words and ideas? Or maybe you've memorized facts for a test, and then weeks later you're not even sure what topic those facts are related to.

In both cases, you may have understood the ideas or concepts by themselves, but not in relation to one another. If you could somehow link the ideas together, you would probably understand them better and remember them longer. This is something a concept map can help you do. A concept map is a visual way of choosing how ideas or concepts fit together. It can help you see the "big picture."

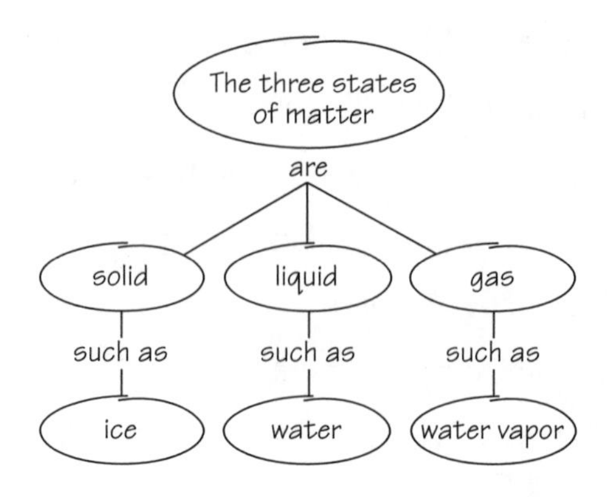


## How to Make a Concept Map

1. Make a list of the main ideas or concepts. It might help to write each concept on its own slip of paper. This will make it easier to rearrange the concepts as many times as you need to before you've made sense of how the concepts are connected. After you've made a few concept maps this way, you can go directly from writing your list to actually making the map.
2. Spread out the slips on a sheet of paper, and arrange the concepts in order from the most general to the most specific. Put the most general concept at the top and circle it. Ask yourself, "How does this concept relate to the remaining concepts?" As you see the relationships, arrange the concepts in order from general to specific.
3. Connect the related concepts with lines.
4. On each line, write an action word or short phrase that shows how the concepts are related.

Look at the concept map on this page and then see if you can make one for the following terms: The Solar System, The Sun, Planets, Star, Earth, Jupiter.

*An answer is provided below, but don't look at it until you try the concept map yourself.*

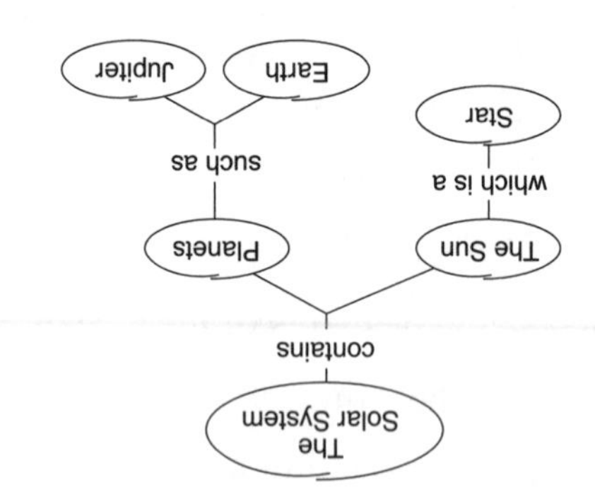

Name ______________________ Class ______________ Date ______________



Chapter 1

# Introduction to Earth Science

## Review

**Choose the best response. Write the letter of that choice in the space provided.**

_____ **1.** The study of the solid earth is called

**a.** geology. **b.** oceanography. **c.** meteorology. **d.** astronomy.

_____ **2.** The earth scientist most likely to study storms is

**a.** a geologist. **b.** an oceanographer.
**c.** a meteorologist. **d.** an astronomer.

_____ **3.** The study of the complex relationships between living things and their environment is called

**a.** geology. **b.** meteorology. **c.** ecology. **d.** astronomy.

_____ **4.** An example of a nonbiodegradable waste product is

**a.** an apple core. **b.** a plastic milk jug.
**c.** a pile of rotting leaves. **d.** an eggshell.

_____ **5.** Usually the first step in scientific problem solving is to

**a.** form a hypothesis. **b.** state the problem.
**c.** gather information. **d.** state a conclusion.

_____ **6.** A possible explanation for a scientific problem is called

**a.** an experiment. **b.** an observation. **c.** a theory. **d.** a hypothesis.

_____ **7.** The development of the meteorite-impact hypothesis began with the observation of

**a.** blue shift in the spectra of stars. **b.** red shift in the spectra of stars.
**c.** background radiation. **d.** iridium in earth rocks.

_____ **8.** A statement that consistently and correctly describes some natural phenomenon is a scientific

**a.** hypothesis. **b.** observation. **c.** law. **d.** control.

_____ **9.** The apparent change in the wavelengths of a moving energy source is called

**a.** the big bang. **b.** the Doppler effect.
**c.** the spectrum. **d.** background radiation.

## Chapter 1

**Choose the best response. Write the letter of that choice in the space provided.**

_____ **10.** Scientists have found that as a light source moves toward a stationary observer, the wavelengths of the light source appear

**a.** longer. **b.** shorter. **c.** higher. **d.** lower.

_____ **11.** The big bang theory states that the galaxies in the universe are

**a.** moving away from one another. **b.** moving toward one another.
**c.** remaining stationary. **d.** being bombarded by meteoroids.

_____ **12.** Evidence for the big bang theory includes

**a.** iridium in earth rocks. **b.** deformed quartz particles in earth rocks.
**c.** blue shift in the spectra of galaxies. **d.** red shift in the spectra of galaxies.

## Critical Thinking

**Read each question or statement and answer it in the space provided.**

**1.** A meteorite lands in your backyard. Which earth scientist would you call to study the meteorite? Why?

______________________________________________

______________________________________________

______________________________________________

______________________________________________

______________________________________________

______________________________________________

**2.** A stream that feeds a small pond gradually dries up. How might this change affect the ecosystem of the pond?

______________________________________________

______________________________________________

______________________________________________

______________________________________________

______________________________________________

______________________________________________

## Chapter 1

**Read each question or statement and answer it in the space provided.**

**3.** Some scientists have hypothesized that meteorites have periodically bombarded the earth, causing mass extinctions every 26 million years. How might this hypothesis be tested?

______________________________

______________________________

______________________________

______________________________

______________________________

______________________________

**4.** Imagine that you are on another planet in a galaxy far from the earth. If you used a spectroscope to examine the spectrum of the sun, would you expect to find red shift, blue shift, or no shift at all? Why?

______________________________

______________________________

______________________________

______________________________

______________________________

______________________________

**5.** According to the big bang theory, the original big bang took place about 15 billion years ago. How might scientists have been able to determine this?

______________________________

______________________________

______________________________

______________________________

______________________________

______________________________

## Chapter 1

## Application

**Read each question or statement and answer it in the space provided.**

**1.** You find a yellow rock and wonder if it is gold. How could you apply scientific methods to this problem?

______________________________________________________________________

______________________________________________________________________

______________________________________________________________________

______________________________________________________________________

**2.** A scientist observes that each eruption of a volcano is preceded by a series of small earthquakes. The scientist then makes the following statement: Earthquakes cause volcanic eruptions. Is the scientist's statement a hypothesis or a theory? Why?

______________________________________________________________________

______________________________________________________________________

______________________________________________________________________

______________________________________________________________________

**3.** Construct a **concept map** using 10 of the new terms listed on page 18 of your textbook by making connections that illustrate the relationship among the terms. See page iv of this workbook for instructions on making concept maps.

## Chapter 2

# The Earth in Space

### Review

**Choose the best response. Write the letter of that choice in the space provided.**

_____ **1.** The zone that makes up nearly two-thirds of the earth's mass is the

**a.** crust. **b.** mantle. **c.** core. **d.** lithosphere.

_____ **2.** The boundary between the earth's crust and mantle is called the

**a.** shadow zone. **b.** asthenosphere. **c.** Moho. **d.** magnetosphere.

_____ **3.** Both P waves and S waves can travel through

**a.** liquids and solids. **b.** solids.
**c.** liquids. **d.** gases.

_____ **4.** The possible source of the earth's magnetism is the earth's

**a.** crust. **b.** mantle. **c.** core. **d.** lithosphere.

_____ **5.** The amount of matter in an object is the object's

**a.** mass. **b.** weight. **c.** gravity. **d.** plasticity.

_____ **6.** As the distance from the center of the earth increases, the force of gravity

**a.** decreases. **b.** increases. **c.** stays the same. **d.** doubles.

_____ **7.** The measured weight of an object is slightly less at the equator than it is at the poles because of the earth's

**a.** orbit. **b.** axis. **c.** shape. **d.** tilt.

_____ **8.** The point closest to the sun in the earth's orbit is called

**a.** apogee. **b.** perigee. **c.** aphelion. **d.** perihelion.

_____ **9.** At noon on the winter solstice, the sun's vertical rays strike the earth along the

**a.** Tropic of Cancer. **b.** Tropic of Capricorn.
**c.** equator. **d.** North Pole.

_____ **10.** At noon on the vernal equinox, the sun's vertical rays strike the earth along the

**a.** Tropic of Cancer. **b.** Tropic of Capricorn.
**c.** equator. **d.** North Pole.

## Chapter 2

**Choose the best response. Write the letter of that choice in the space provided.**

_____ **11.** When the sun's rays reach their highest angle in the Northern Hemisphere, the season there is

**a.** spring. **b.** summer. **c.** fall. **d.** winter.

_____ **12.** The wobbling motion made by the earth's axis as it turns in space is called

**a.** precession. **b.** gravity. **c.** plasticity. **d.** equinox.

_____ **13.** A person crossing the International Date Line gains or loses

**a.** 2 hours. **b.** 8 hours. **c.** 12 hours. **d.** 24 hours.

_____ **14.** A satellite in geosynchronous orbit is always directly above the

**a.** equator. **b.** North Pole.
**c.** South Pole. **d.** International Date Line.

_____ **15.** Landsat provides information about the

**a.** earth's surface. **b.** moon's surface.
**c.** sun's surface. **d.** stars.

## Critical Thinking

**Read each question or statement and answer it in the space provided.**

**1.** Is a hard-boiled egg a good model of the earth's different zones? Why or why not?

________________________________________________
________________________________________________
________________________________________________
________________________________________________

**2.** Explain why the weight of an object might increase when the object moves from point A to point B on the earth's surface.

________________________________________________
________________________________________________
________________________________________________
________________________________________________

## Chapter 2

**Read each question or statement and answer it in the space provided.**

**3.** Although the earth's orbit brings it closest to the sun in January, the Northern Hemisphere is having winter at that time of the year. Explain.

**4.** If the earth ceased rotating as it revolved around the sun, how would periods of daylight and surface temperatures on the earth be affected?

**5.** Suppose the earth's rotation slowed to one rotation every 48 hours. How might that change affect time-keeping systems?

## Chapter 2

## Application

**Read each question or statement and answer it in the space provided.**

**1.** A scientist proposes the hypothesis that the moon has a liquid core. How might this hypothesis be tested?

______________________________________________

______________________________________________

______________________________________________

______________________________________________

______________________________________________

**2.** If scientists discovered that the earth's magnetic field had weakened substantially, what might they suspect to be the cause of this change?

______________________________________________

______________________________________________

______________________________________________

______________________________________________

______________________________________________

**3.** A group of scientists is planning to launch a satellite into orbit around the earth. The satellite will be used to survey the entire earth's surface in search of oil deposits. Which type of orbit would you recommend for such a satellite? Why?

______________________________________________

______________________________________________

______________________________________________

______________________________________________

______________________________________________

Name ______________________ Class ______________ Date ______________



Chapter 3

# Models of the Earth

## Review

**Choose the best response. Write the letter of that choice in the space provided.**

_____ **1.** A point whose latitude is 0° is located on the

**a.** North Pole. **b.** South Pole. **c.** equator. **d.** prime meridian.

_____ **2.** One degree of latitude equals

**a.** 1/90 the earth's circumference. **b.** 1/100 the earth's circumference.
**c.** 1/360 the earth's circumference. **d.** 1/720 the earth's circumference.

_____ **3.** A point whose longitude is 0° is located on the

**a.** North Pole. **b.** South Pole. **c.** equator. **d.** prime meridian.

_____ **4.** A point halfway between the equator and the South Pole has a latitude of

**a.** 45° N. **b.** 45° S. **c.** 45° E. **d.** 45° W.

_____ **5.** The distance in degrees east or west of the prime meridian is

**a.** latitude. **b.** longitude. **c.** declination. **d.** projection.

_____ **6.** The distance covered by a degree of longitude

**a.** is 1/180 the earth's circumference. **b.** is 1/360 the earth's circumference.
**c.** increases as you approach the poles. **d.** decreases as you approach the poles.

_____ **7.** The needle of a magnetic compass points toward the

**a.** geomagnetic pole. **b.** geographic pole.
**c.** parallels. **d.** meridians.

_____ **8.** In the Northern Hemisphere, a declination of 10° E indicates that the compass needle points 10° east of the

**a.** geomagnetic North Pole. **b.** geographic North Pole.
**c.** equator. **d.** prime meridian.

_____ **9.** On a Mercator projection, distortion is greatest near the

**a.** poles. **b.** great circles. **c.** meridians. **d.** parallels.

Name ______________________ Class ______________ Date ______________



## Chapter 3

**Choose the best response. Write the letter of that choice in the space provided.**

_____ **10.** Compass directions are shown as straight lines on a

**a.** gnomonic projection. **b.** conic projection.
**c.** Mercator projection. **d.** polyconic projection.

_____ **11.** The shortest distance between any two points on the globe is along

**a.** the equator. **b.** a line of latitude.
**c.** the prime meridian. **d.** a great circle.

_____ **12.** A navigator can find the shortest distance between two points by drawing a straight line between any two points on a

**a.** Mercator projection. **b.** gnomonic projection.
**c.** conic projection. **d.** polyconic projection.

_____ **13.** The relationship between distance on a map and actual distance on the earth is called the

**a.** legend. **b.** scale. **c.** elevation. **d.** relief.

_____ **14.** If 1 m on a map equals 1 km on the earth, the fractional scale would be written

**a.** 1:1. **b.** 1:10. **c.** 1:100. **d.** 1:1,000.

_____ **15.** On a topographic map, elevation is shown by means of

**a.** great circles. **b.** contour lines. **c.** verbal scale. **d.** fractional scale.

_____ **16.** Closely spaced contour lines indicate a

**a.** gradual slope. **b.** flat area. **c.** steep slope. **d.** valley.

## Critical Thinking

**Read each question or statement and answer it in the space provided.**

**1.** What is wrong with the following location: 135° N, 185° E?

______________________________________________

______________________________________________

______________________________________________

______________________________________________

______________________________________________

## Chapter 3

**Read each question or statement and answer it in the space provided.**

**2.** As you move from point A to point B in the Northern Hemisphere, the length of a degree of longitude progressively decreases. In which direction are you moving?

**3.** Imagine you are at a location where the magnetic declination is 0°. Describe your position in relation to magnetic north and true north.

**4.** You examine a topographic map on which the contour interval is 100 m. In general, what type of terrain is shown on the map?

**5.** Selecting from the list of new terms on the previous page, which one term would most likely be found at the top of a concept map designed for this chapter? Explain.

## Chapter 3

## Application

**Read each question or statement and answer it in the space provided.**

**1.** One expedition is preparing to explore the South Pole; another is preparing to explore the equator. To which expedition would you recommend the Mercator projection? Explain why.

______________________________________________

______________________________________________

______________________________________________

______________________________________________

**2.** A cartographer has to draw one map for use in three different countries that do not share a common unit of measure. Which type of scale should this mapmaker use? Why?

______________________________________________

______________________________________________

______________________________________________

______________________________________________

**3.** You are using a topographic map to plan a hike. Along path A, the contour lines are widely spaced. Along path B, the contour lines are almost touching. Which path would probably be easier and safer? Why?

______________________________________________

______________________________________________

______________________________________________

______________________________________________

**4.** How could you use contour lines on a topographic map to help you locate the source of a river?

______________________________________________

______________________________________________

______________________________________________

______________________________________________

Chapter 4

# Plate Tectonics

## Review

**Choose the best response. Write the letter of that choice in the space provided.**

_____ **1.** The German scientist Alfred Wegener proposed the existence of a huge landmass called

**a.** Panthalassa. **b.** rift valley. **c.** *Mesosaurus.* **d.** Pangaea.

_____ **2.** Support for Wegener's hypothesis of continental drift includes evidence of changes in

**a.** climatic patterns. **b.** Panthalassa.
**c.** terranes. **d.** subduction.

_____ **3.** New ocean floor is constantly being produced through the process known as

**a.** subduction. **b.** continental drift.
**c.** seafloor spreading. **d.** terranes.

_____ **4.** An underwater mountain chain formed where new crust is created by seafloor spreading is called a

**a.** divergent boundary. **b.** subduction zone.
**c.** mid-ocean ridge. **d.** convergent boundary.

_____ **5.** The term *tectonics* comes from a Greek word meaning

**a.** "movement." **b.** "plate." **c.** "continent." **d.** "construction."

_____ **6.** The layer of mantle with plastic rock that underlies the plates is called the

**a.** lithosphere. **b.** asthenosphere. **c.** oceanic crust. **d.** terrane.

_____ **7.** To date, scientists have identified approximately

**a.** 5 plates. **b.** 30 plates. **c.** 15 plates. **d.** 50 plates.

_____ **8.** Two plates moving away from each other form a

**a.** transform boundary. **b.** convergent boundary.
**c.** fracture. **d.** divergent boundary.

_____ **9.** The collision of one lithospheric plate with another forms a

**a.** convergent boundary. **b.** transform boundary.
**c.** rift valley. **d.** divergent boundary.

## Chapter 4

**Choose the best response. Write the letter of that choice in the space provided.**

_____ **10.** The region along lithospheric plate boundaries where one plate is moved beneath another is called a

**a.** rift valley.
**b.** transform boundary.
**c.** subduction zone.
**d.** convergent boundary.

_____ **11.** Two plates grind past each other at a

**a.** transform boundary.
**b.** convergent boundary.
**c.** subduction zone.
**d.** divergent boundary.

_____ **12.** Convection occurs because heated material becomes

**a.** less dense and rises.
**b.** more dense and rises.
**c.** more dense and sinks.
**d.** less dense and sinks.

_____ **13.** Scientists think that the upwelling of mantle material at mid-ocean ridges is caused by the motion of lithospheric plates and comes from

**a.** the lithosphere.
**b.** the asthenosphere.
**c.** terranes.
**d.** rift valleys.

_____ **14.** The parts of the continents that are made up of formerly separate pieces of lithosphere are called

**a.** terranes.
**b.** plates.
**c.** continental crust.
**d.** oceanic crust.

## Critical Thinking

**Read each question or statement and answer it in the space provided.**

**1.** In what ways might the concept of continental drift be compared to a jigsaw puzzle?

______________________________________________________________

______________________________________________________________

______________________________________________________________

______________________________________________________________

______________________________________________________________

______________________________________________________________

## Chapter 4

**Read each question or statement and answer it in the space provided.**

**2.** If Alfred Wegener had found identical fossil remains of plants and animals that had lived no more than 10 million years ago in both eastern Brazil and western Africa, what might he have concluded about the breakup of Pangaea?

______

______

______

______

______

______

**3.** Assume that the total surface area of the earth is not changing. If new material is being added to the earth's crust at one boundary, what would you expect to find happening at another boundary?

______

______

______

______

______

______

**4.** Explain the following statement: Due to the process of seafloor spreading, the ocean floor is constantly renewing itself.

______

______

______

______

______

______

## Chapter 4

## Application

**Read each question or statement and answer it in the space provided.**

**1.** Explain the role of technology in the progression from the hypothesis of continental drift to the theory of plate tectonics.

______________________________

______________________________

______________________________

______________________________

**2.** If you wanted to prove that a microplate terrane had been scraped onto the North American plate, what kind of evidence would you search for?

______________________________

______________________________

______________________________

______________________________

**3.** Construct a **concept map** using 10 of the new terms listed on page 78 of your textbook by making connections that illustrate the relationship among the terms.

Name ____________________ Class ____________ Date ____________



Chapter 5

# Deformation of the Crust

## Review

**Choose the best response. Write the letter of that choice in the space provided.**

_____ **1.** The state of balance between the thickness of the crust and the depth at which it rides on the asthenosphere is called

**a.** stress. **b.** isostasy. **c.** strain. **d.** shearing.

_____ **2.** The increasing weight of mountains causes the crust to

**a.** sink. **b.** fold. **c.** rise. **d.** fracture.

_____ **3.** The force that changes the shape and volume of rocks is

**a.** footwall. **b.** isostasy. **c.** rising. **d.** stress.

_____ **4.** The type of stress that squeezes rock together is

**a.** compression. **b.** tension. **c.** shearing. **d.** faulting.

_____ **5.** The type of stress that pulls rocks apart, making them thinner, is

**a.** folding. **b.** compression. **c.** tension. **d.** isostasy.

_____ **6.** Shearing

**a.** bends, twists, or breaks rocks. **b.** squeezes rock together.
**c.** causes rock to melt. **d.** pulls rock apart.

_____ **7.** High pressure and high temperature will cause rocks to

**a.** fracture. **b.** adjust. **c.** plateau. **d.** deform.

_____ **8.** Upcurved folds in rock are called

**a.** anticlines. **b.** monoclines. **c.** fractures. **d.** synclines.

_____ **9.** Downcurved folds in rock are called

**a.** fractures. **b.** monoclines. **c.** anticlines. **d.** synclines.

_____ **10.** Folds in which both limbs remain horizontal are called

**a.** monoclines. **b.** fractures. **c.** synclines. **d.** anticlines.

Name ______________________ Class ______________ Date ____________



## Chapter 5

**Choose the best response. Write the letter of that choice in the space provided.**

_____ **11.** When no movement occurs along the sides of a break in a rock structure, the break is called a

**a.** normal fault. **b.** fracture. **c.** fold. **d.** hanging wall.

_____ **12.** When a fault is not vertical, the rock above the fault plane makes up the

**a.** tension. **b.** footwall. **c.** hanging wall. **d.** compression.

_____ **13.** A nearly vertical fault in which the rock on either side of the fault plane moves horizontally is called a

**a.** normal fault. **b.** reverse fault. **c.** strike-slip fault. **d.** thrust fault.

_____ **14.** The largest mountain systems are part of still larger systems called

**a.** continental margins. **b.** ranges.
**c.** belts. **d.** synclines.

_____ **15.** Mount St. Helens in Washington State is an example of a

**a.** folded mountain. **b.** volcanic mountain.
**c.** fault-block mountain. **d.** dome mountain.

## Critical Thinking

**Read each question or statement and answer it in the space provided.**

**1.** Suppose glaciers, which are vast fields of slow-moving ice, were to cover much of the earth's surface once again. What would you expect to happen to those parts of the continents that were covered by ice? Explain.

______________________________________________

______________________________________________

______________________________________________

______________________________________________

______________________________________________

______________________________________________

## Chapter 5

**Read each question or statement and answer it in the space provided.**

**2.** When the Indian plate collided with the Eurasian plate, producing the Himalaya Mountains, which type of stress most likely occurred? Which type of stress is most likely occurring along the Mid-Atlantic Ridge? Which type of stress would you expect to find along the San Andreas Fault? Use your knowledge of stress and plate tectonics to explain your answers.

______________________________

______________________________

______________________________

______________________________

______________________________

______________________________

**3.** If the force that is causing a rock to be slightly deformed begins to ease, what might happen to the rock? What would happen if the force causing the deformation became greater?

______________________________

______________________________

______________________________

______________________________

______________________________

______________________________

**4.** Why do you suppose dome mountains do not become volcanic mountains?

______________________________

______________________________

______________________________

______________________________

______________________________

______________________________

## Chapter 5

## Application

**Read each question or statement and answer it in the space provided.**

**1.** Suppose that a new highway is being planned. This proposed road would intersect a transform boundary. What would happen to the highway if a strike-slip fault existed along the boundary? Why?

______________________________________________________________________

______________________________________________________________________

______________________________________________________________________

______________________________________________________________________

**2.** A geologist discovers that part of a mountain range along the west coast of the United States contains the fossil remains of animals that do not match any other fossils from North America. What is the most likely explanation for this phenomenon?

______________________________________________________________________

______________________________________________________________________

______________________________________________________________________

______________________________________________________________________

**3.** Construct a **concept map** that illustrates the relationship between crustal deformation and types of mountains.

Chapter 6

# Earthquakes

## Review

**Choose the best response. Write the letter of that choice in the space provided.**

_____ **1.** Vibrations in the earth caused by the sudden movement of rock are called

**a.** epicenters. **b.** earthquakes. **c.** faults. **d.** tsunamis.

_____ **2.** The elastic rebound theory states that as a rock becomes stressed, it first

**a.** deforms. **b.** melts. **c.** breaks. **d.** shifts position.

_____ **3.** The point along a fault where an earthquake begins is called the

**a.** fracture. **b.** epicenter. **c.** gap. **d.** focus.

_____ **4.** The point on the earth's surface directly above the point where an earthquake occurs is called the

**a.** focus. **b.** epicenter. **c.** fracture. **d.** fault.

_____ **5.** A characteristic of earthquakes that causes the most severe damage is

**a.** a deep focus. **b.** an intermediate focus.
**c.** a shallow focus. **d.** a deep epicenter.

_____ **6.** Most severe earthquakes occur

**a.** in mountains. **b.** along major rivers.
**c.** at plate boundaries. **d.** in the middle of plates.

_____ **7.** The boundary of the Pacific plate scrapes against that of the North American plate and forms

**a.** a single fault. **b.** a subduction zone.
**c.** a volcano. **d.** a fault zone.

_____ **8.** P waves travel through

**a.** solids only. **b.** liquids and gases only.
**c.** both solids and liquids. **d.** liquids only.

_____ **9.** S waves cannot pass through

**a.** solids. **b.** the mantle.
**c.** the earth's outer core. **d.** the asthenosphere.

## Chapter 6

**Choose the best response. Write the letter of that choice in the space provided.**

_____ **10.** By analyzing the difference in the time it takes for P waves and S waves to arrive at a seismograph station, scientists can determine an earthquake's

**a.** epicenter. **b.** surface waves. **c.** fault zone. **d.** intensity.

_____ **11.** The Richter scale expresses an earthquake's

**a.** magnitude. **b.** location. **c.** duration. **d.** depth.

_____ **12.** Most injuries during earthquakes are caused by

**a.** the collapse of buildings.
**b.** cracks in the earth's surface.
**c.** the vibration of S waves.
**d.** the vibration of P waves.

_____ **13.** If an earthquake strikes while you are in a car, you should

**a.** continue driving.
**b.** get out of the car.
**c.** park the car under a bridge.
**d.** stop the car in a clear space and remain in the car.

_____ **14.** An earthquake is frequently preceded by

**a.** a temporary change in the speed of local P waves.
**b.** a temporary change in the speed of the surface waves.
**c.** landslides.
**d.** tsunamis.

## Critical Thinking

**Read each question or statement and answer it in the space provided.**

**1.** Describe how the elastic rebound theory explains the generation of seismic waves.

## Chapter 6

**Read each question or statement and answer it in the space provided.**

**2.** If a seismograph station measures P waves but no S waves from an earthquake, what can you conclude about the earthquake's location?

______________________________________________

______________________________________________

______________________________________________

______________________________________________

______________________________________________

______________________________________________

**3.** Two cities are struck by the same earthquake. The cities are the same size, are built on the same type of ground, and have the same types of buildings. The city in which the quake measured IV on the Mercalli scale suffered $1 million in damage. The city in which the quake measured VI on the Mercalli scale suffered $50 million in damage. What might account for this great difference in damage costs?

______________________________________________

______________________________________________

______________________________________________

______________________________________________

______________________________________________

______________________________________________

**4.** Would an earthquake in the Colorado Rockies be likely to form a tsunami? Explain why.

______________________________________________

______________________________________________

______________________________________________

______________________________________________

______________________________________________

______________________________________________

Name ______________________ Class ______________ Date ______________



## Chapter 6

## Application

**Read each question or statement and answer it in the space provided.**

**1.** You are going to choose a building site for a home. You would like a high place with a view, but you are concerned about earthquakes. What information do you need to make an informed decision about the site?

________________________________________

________________________________________

________________________________________

________________________________________

________________________________________

**2.** Why is it wise to stand in a doorway when an earthquake strikes?

________________________________________

________________________________________

________________________________________

________________________________________

________________________________________

**3.** Imagine that you are monitoring a seismograph station along the San Andreas Fault. An earthquake occurs in Mexico, and you notice that the P waves you are recording from that quake have a velocity that is less than normal. What does this tell you about the area around your seismograph station?

________________________________________

________________________________________

________________________________________

________________________________________

________________________________________

# Chapter 7

# Volcanoes

## Review

**Choose the best response. Write the letter of that choice in the space provided.**

_____ **1.** Factors that allow magma to push its way upward include temperature and

**a.** color. **b.** density. **c.** crust. **d.** thickness.

_____ **2.** Activity caused by the movement of magma is called

**a.** extraterrestrial. **b.** pyroclastic. **c.** volcanism. **d.** subduction.

_____ **3.** The belt of volcanoes that encircles the Pacific Ocean is called

**a.** the subduction zone. **b.** an island arc.
**c.** a hot spot. **d.** the Pacific Ring of Fire.

_____ **4.** Island arcs are formed by the collision of

**a.** two plates with continental crust at their edges.
**b.** two calderas.
**c.** two volcanic bombs.
**d.** two plates with oceanic crust at their edges.

_____ **5.** Areas of volcanism within plates are called

**a.** hot spots. **b.** calderas. **c.** cones. **d.** fissures.

_____ **6.** Lava that breaks into jagged chunks when it is subjected to rapid cooling is called

**a.** aa. **b.** pahoehoe. **c.** pillow lava. **d.** felsic lava.

_____ **7.** Explosive volcanic eruptions result from

**a.** mafic lava. **b.** aa lava. **c.** felsic lava. **d.** pahoehoe lava.

_____ **8.** Pyroclastic materials that form into rounded or spindle shapes as they fly through the air are called

**a.** ash. **b.** lapilli.
**c.** volcanic bombs. **d.** volcanic blocks.

_____ **9.** The Hawaiian Islands are formed from

**a.** shield volcanoes. **b.** cinder cones.
**c.** composite volcanoes. **d.** calderas.

## Chapter 7

**Choose the best response. Write the letter of that choice in the space provided.**

_____ **10.** A cone formed only by solid fragments built up around a volcanic opening is a

**a.** shield volcano. **b.** cinder cone.
**c.** composite volcano. **d.** stratovolcano.

_____ **11.** The depression that results when a cone collapses into an empty magma chamber is a

**a.** crater. **b.** vent. **c.** caldera. **d.** fissure.

_____ **12.** Shortly before a volcano erupts, magma may cause its surface to

**a.** bulge out. **b.** cave in. **c.** get darker. **d.** melt.

_____ **13.** Scientists have discovered that before an eruption, earthquakes

**a.** completely stop. **b.** increase in number.
**c.** bear no relation to volcanism. **d.** decrease in number.

_____ **14.** Olympus Mons, the largest known volcano in the solar system, is found on

**a.** the moon. **b.** Io. **c.** Mars. **d.** the earth.

_____ **15.** The material ejected from volcanoes on Io is probably

**a.** sulfur and sulfur dioxide. **b.** basalt.
**c.** granite. **d.** lapilli.

## Critical Thinking

**Read each question or statement and answer it in the space provided.**

**1.** Most lava that forms on the earth's surface goes unnoticed and unobserved. Why does this happen?

______________________________________________

______________________________________________

______________________________________________

______________________________________________

______________________________________________

______________________________________________

## Chapter 7

**Read each question or statement and answer it in the space provided.**

**2.** Can you assume that every mountain located along the edge of a continent is a volcano? Why or why not?

**3.** Why does felsic lava tend to produce composite volcanoes and cinder cones rather than shield volcanoes?

**4.** How might geologists distinguish an impact crater on earth, such as Meteor Crater in Arizona, from a volcanic feature?

## Application

**Read each question or statement and answer it in the space provided.**

**1.** While exploring a volcano that has been dormant, you observe volcanic ash first and lapilli later. Are you more likely to be moving toward or away from the volcanic opening? Explain your answer.

## Chapter 7

**Read each question or statement and answer it in the space provided.**

**2.** If you see a steep volcanic cone that is only 300 m high, what can you assume about the type of cone and its composition?

______________________________________________

______________________________________________

______________________________________________

______________________________________________

**3.** To predict a volcanic eruption, what kinds of information would you seek?

______________________________________________

______________________________________________

______________________________________________

______________________________________________

**4.** Construct a **concept map** using as many new terms as possible from the list on page 128 of your textbook to illustrate the concepts in this chapter.

Chapter 8

# Earth Chemistry

## Review

**Choose the best response. Write the letter of that choice in the space provided.**

_____ **1.** Color and hardness are examples of an element's
- **a.** physical properties.
- **b.** chemical properties.
- **c.** atomic structure.
- **d.** molecular properties.

_____ **2.** A substance that cannot be broken down into a simpler form by ordinary chemical means is
- **a.** a mixture.
- **b.** a gas.
- **c.** an element.
- **d.** a compound.

_____ **3.** The smallest unit of an element is
- **a.** a molecule.
- **b.** an atom.
- **c.** an ion.
- **d.** an electron.

_____ **4.** Particles in atoms that do not carry an electrical charge are called
- **a.** neutrons.
- **b.** nuclei.
- **c.** protons.
- **d.** ions.

_____ **5.** The number of protons in the nucleus indicates the atom's
- **a.** mass number.
- **b.** electrical charges.
- **c.** isotope.
- **d.** atomic number.

_____ **6.** The mass number of an atom is equal to its
- **a.** total number of protons.
- **b.** total number of electrons and protons.
- **c.** total number of neutrons and protons.
- **d.** total number of neutrons.

_____ **7.** Atoms of the same element that differ in mass are
- **a.** ions.
- **b.** isotopes.
- **c.** neutrons.
- **d.** molecules.

_____ **8.** A material with a definite shape and volume is a
- **a.** compound.
- **b.** liquid.
- **c.** gas.
- **d.** solid.

_____ **9.** A liquid does not have a definite
- **a.** shape.
- **b.** volume.
- **c.** chemical formula.
- **d.** mass.

_____ **10.** If a gas is not confined, the space between its particles will
- **a.** decrease slowly.
- **b.** decrease rapidly.
- **c.** increase.
- **d.** not change.

## Chapter 8

**Choose the best response. Write the letter of that choice in the space provided.**

_____ **11.** Atoms of two or more elements that are chemically united form

**a.** a mixture. **b.** a nucleus. **c.** an ion. **d.** a compound.

_____ **12.** An atom does not easily lose or gain electrons if it has

**a.** many protons. **b.** a filled outer energy level.
**c.** many energy levels. **d.** few neutrons.

_____ **13.** A molecule of water, or $H_2O$, has one atom of

**a.** hydrogen. **b.** helium. **c.** oxygen. **d.** osmium.

_____ **14.** A material that contains two or more substances that are not chemically combined is

**a.** a mixture. **b.** a compound. **c.** an ion. **d.** a molecule.

## Critical Thinking

**Read each question or statement and answer it in the space provided.**

**1.** Oxygen combines with hydrogen to form water. Is this process a result of the physical or chemical properties of oxygen?

______________________________________________

______________________________________________

**2.** What distinguishes an atom of one element from atoms of all other elements?

______________________________________________

______________________________________________

**3.** Why do isotopes of an element have different mass numbers?

______________________________________________

______________________________________________

______________________________________________

______________________________________________

## Chapter 8

**Read each question or statement and answer it in the space provided.**

**4.** The mercury in a thermometer has a volume that varies with temperature. It takes the shape of the glass tube that holds it. Is the mercury in a thermometer a solid, a liquid, or a gas?

**5.** Calcium chloride is an ionic compound. Carbon dioxide is a covalent compound. Which of these compounds would you expect to have a lower melting point? Explain your answer.

**6.** Is a diatomic molecule more likely to be held together by a covalent bond or an ionic bond? Explain why you think this is so.

**7.** What happens to the chemical properties of a substance when it becomes part of a mixture?

Name ______________________ Class ______________ Date ______________



## Chapter 8

## Application

**Read each question or statement and answer it in the space provided.**

**1.** How many neutrons does a potassium atom have if its atomic number is 19 and its mass number is 39?

______________________________________________

______________________________________________

______________________________________________

______________________________________________

**2.** The atomic number of calcium is 20, and the atomic number of copper is 29. Which has more electrons, a calcium atom or a copper atom? How do you know?

______________________________________________

______________________________________________

______________________________________________

______________________________________________

**3.** A helium atom has two electrons in its first and only energy level. Would you predict that helium easily forms compounds with other elements? Why or why not?

______________________________________________

______________________________________________

______________________________________________

______________________________________________

**4.** The chemical formula of glucose (sugar) is $C_6H_{12}O_6$. How many atoms of each element does a molecule of glucose contain?

______________________________________________

______________________________________________

______________________________________________

______________________________________________

Chapter 9

# Minerals of the Earth's Crust

## Review

**Choose the best response. Write the letter of that choice in the space provided.**

_____ **1.** A natural, inorganic, crystalline solid with a characteristic chemical composition is called

**a.** an atom. **b.** a gemstone. **c.** a mineral. **d.** a tetrahedron.

_____ **2.** Minerals that contain silicon and oxygen are

**a.** sulfide minerals. **b.** sulfate minerals.
**c.** ores. **d.** silicate minerals.

_____ **3.** The most common silicate minerals are the

**a.** feldspars. **b.** halides. **c.** carbonates. **d.** sulfates.

_____ **4.** Ninety-six percent of the earth's crust is made up of

**a.** sulfur and lead. **b.** silicate minerals.
**c.** copper and aluminum. **d.** nonsilicate minerals.

_____ **5.** The basic structural units of all silicate minerals consist of

**a.** tetrahedral frameworks. **b.** silicon-oxygen tetrahedra.
**c.** single chains. **d.** double chains.

_____ **6.** An example of a mineral with a basic structure consisting of single tetrahedra linked by atoms of other elements is

**a.** mica. **b.** olivine. **c.** quartz. **d.** feldspar.

_____ **7.** When two single chains of tetrahedra bond to each other, the result is called a

**a.** single-chain silicate. **b.** sheet silicate.
**c.** framework silicate. **d.** double-chain silicate.

_____ **8.** The appearance of the light reflected from the surface of a mineral is called

**a.** color. **b.** streak. **c.** luster. **d.** fluorescence.

_____ **9.** The words *waxy*, *pearly*, and *dull* describe a mineral's

**a.** luster. **b.** hardness. **c.** streak. **d.** fluorescence.

## Chapter 9

**Choose the best response. Write the letter of that choice in the space provided.**

_____ **10.** The words *uneven* and *splintery* describe a mineral's

**a.** cleavage. **b.** fracture. **c.** hardness. **d.** luster.

_____ **11.** Mohs scale is used in measuring a mineral's

**a.** hardness. **b.** cleavage. **c.** color. **d.** luster.

_____ **12.** The ratio of the mass of a mineral to its volume is the mineral's

**a.** atomic weight. **b.** density. **c.** mass. **d.** weight.

_____ **13.** The needles of the first magnetic compasses used in navigation were made of the magnetic mineral

**a.** iron pyrite. **b.** silver. **c.** cinnabar. **d.** lodestone.

_____ **14.** When calcite absorbs ultraviolet light and gives off red light, it is displaying the property of

**a.** radioactivity. **b.** double refraction.
**c.** magnetism. **d.** fluorescence.

_____ **15.** A mineral that is radioactive probably contains the element

**a.** uranium. **b.** silicon. **c.** fluorine. **d.** calcium.

_____ **16.** Double refraction is a distinctive property of crystals of

**a.** mica. **b.** feldspar. **c.** calcite. **d.** galena.

## Critical Thinking

**Read each question or statement and answer it in the space provided.**

**1.** Natural gas is a substance that occurs naturally in the earth's crust. Is it a mineral? Explain how you know.

_______________________________________________

_______________________________________________

_______________________________________________

_______________________________________________

_______________________________________________

_______________________________________________

Name ______________________ Class ______________ Date ______________

MODERN EARTH SCIENCE

## Chapter 9

**Read each question or statement and answer it in the space provided.**

**2.** Which of the following are you most likely to find in the earth's crust: the silicates feldspar and quartz or the nonsilicates copper and iron? Explain your answer.

______________________________________________

______________________________________________

______________________________________________

______________________________________________

**3.** Which, if any, of the following mineral groups contain silicon: carbonates, halides, or sulfates? Explain how you know.

______________________________________________

______________________________________________

______________________________________________

______________________________________________

**4.** Describe the tetrahedral arrangement of olivine.

______________________________________________

______________________________________________

______________________________________________

______________________________________________

**5.** Can you determine conclusively that an unknown substance contains magnetite using only a magnet? Explain.

______________________________________________

______________________________________________

______________________________________________

______________________________________________

## Chapter 9

## Application

**Read each question or statement and answer it in the space provided.**

**1.** Why is it difficult to identify a mineral simply by its color?

______________________________________________

______________________________________________

______________________________________________

**2.** Iron pyrite ($FeS_2$) is called *fool's gold* because it looks very much like gold. What simple test could you use to determine whether a mineral sample is gold or pyrite? Explain what it would show.

______________________________________________

______________________________________________

______________________________________________

**3.** A mineral sample has a mass of 51 g and a volume of 15 $cm^3$. What is the density of the mineral sample?

______________________________________________

**4.** Construct a **concept map** starting with the term *minerals* and using as many new terms as possible from the list on page 170 of your textbook. Make connections that illustrate the classification of minerals.

Chapter 10

# Rocks

## Review

**Choose the best response. Write the letter of that choice in the space provided.**

_____ **1.** Rock that forms from magma is called

**a.** igneous. **b.** metamorphic. **c.** sedimentary. **d.** clastic.

_____ **2.** The process in which rock changes from one type to another and back again is called

**a.** a rock family. **b.** the rock cycle.
**c.** contact metamorphism. **d.** foliation.

_____ **3.** Intrusive igneous rocks are characterized by a coarse-grained texture because they contain

**a.** heavy elements. **b.** small crystals. **c.** large crystals. **d.** fragments of different sizes and shapes.

_____ **4.** Light-colored igneous rocks are part of the

**a.** basalt family. **b.** intermediate family.
**c.** felsic family. **d.** mafic family.

_____ **5.** Magma that solidifies underground forms rock masses that are known as

**a.** extrusions. **b.** volcanic cones. **c.** lava plateaus. **d.** intrusions.

_____ **6.** One example of an extrusion is a

**a.** stock. **b.** dike. **c.** batholith. **d.** lava plateau.

_____ **7.** Sedimentary rock formed from rock fragments is called

**a.** organic. **b.** chemical. **c.** clastic. **d.** granite.

_____ **8.** One example of a chemical sedimentary rock is

**a.** evaporites. **b.** coal. **c.** gneiss. **d.** breccia.

_____ **9.** Contact metamorphism is a result of

**a.** plate movement. **b.** hot magma. **c.** sedimentation. **d.** foliation.

_____ **10.** Regional metamorphism is a result of

**a.** plate movement. **b.** hot magma. **c.** cementation. **d.** compaction.

## Chapter 10

**Choose the best response. Write the letter of that choice in the space provided.**

_____ **11.** The splitting of slate into flat layers illustrates its

**a.** contact metamorphism. **b.** formation.
**c.** sedimentation. **d.** foliation.

## Critical Thinking

**Read each question or statement and answer it in the space provided.**

**1.** What type of rock will be formed from a sedimentary rock that comes under extreme pressure and heat but does not melt? Explain your answer.

______________________________________________________________________

______________________________________________________________________

______________________________________________________________________

______________________________________________________________________

______________________________________________________________________

**2.** Explain how metamorphic rock can change into either of the other two types of rock through the rock cycle.

______________________________________________________________________

______________________________________________________________________

______________________________________________________________________

______________________________________________________________________

______________________________________________________________________

**3.** A certain rock is made up mostly of plagioclase feldspar and pyroxene minerals. It also includes olivine and hornblende. Will the rock have a light or dark coloring? Explain your answer.

______________________________________________________________________

______________________________________________________________________

______________________________________________________________________

______________________________________________________________________

______________________________________________________________________

## Chapter 10

**Read each question or statement and answer it in the space provided.**

**4.** Some of the powdery rock found on the moon serves as the cementing agent for sedimentary moon rocks. What type of sedimentary rocks are these? How do you know?

______________________________________________

______________________________________________

______________________________________________

______________________________________________

______________________________________________

______________________________________________

**5.** Imagine that you have found a piece of limestone, a sedimentary rock, with strange-shaped lumps on it. Will the lumps have the same composition as the limestone? Explain your answer.

______________________________________________

______________________________________________

______________________________________________

______________________________________________

______________________________________________

______________________________________________

**6.** Which would be easier to break, the foliated rock slate or the nonfoliated rock quartzite? Explain your answer.

______________________________________________

______________________________________________

______________________________________________

______________________________________________

______________________________________________

______________________________________________

## Chapter 10

## Application

**Read each question or statement and answer it in the space provided.**

**1.** Suppose you found an igneous rock with a coarse texture. Would the magma that formed the rock have cooled slowly or quickly? Explain how you know.

**2.** There is a huge batholith in the northwestern part of Idaho. What can you say about the landscape in that area? Explain your answer.

**3.** If you know that a certain area in South Dakota has a number of laccoliths, what might you expect the landscape to look like?

**4.** The Himalayan mountains are located on a boundary between two colliding tectonic plates. Would most of the metamorphic rock in that area occur in small patches or wide regions? How do you know?

Chapter 11

# Resources and Energy

## Review

**Choose the best response. Write the letter of that choice in the space provided.**

_____ **1.** Metals are known to

**a.** have a dull surface. **b.** provide fuel.
**c.** conduct heat and electricity well. **d.** be found in shale.

_____ **2.** Aluminum can be taken out of a rock called bauxite, which is

**a.** an ore. **b.** an energy source.
**c.** a renewable resource. **d.** a fossil fuel.

_____ **3.** Hot mineral solutions that spread through cracks in rock form bands called

**a.** strata. **b.** crystals. **c.** veins. **d.** silicates.

_____ **4.** Energy resources that have formed from the remains of living things are called

**a.** minerals. **b.** metals. **c.** gemstones. **d.** fossil fuels.

_____ **5.** At the top of an oil reservoir is a layer of

**a.** coal. **b.** cap rock. **c.** peat. **d.** water.

_____ **6.** Plastics, synthetic fabrics, and synthetic rubber are composed of chemicals derived from

**a.** anthracite. **b.** peat. **c.** petroleum. **d.** shale.

_____ **7.** The most abundant fossil fuel is

**a.** coal. **b.** petroleum. **c.** natural gas. **d.** shale.

_____ **8.** One problem caused by the strip mining of coal is

**a.** increased rainfall.
**b.** loss of soil.
**c.** the release of sulfur dioxide into the air.
**d.** the drying up of rivers.

_____ **9.** The splitting of the nucleus of an atom to produce energy is called

**a.** geothermal energy. **b.** nuclear fission.
**c.** nuclear fusion. **d.** hydroelectric power.

Name ________________ Class ________ Date ________



## Chapter 11

**Choose the best response. Write the letter of that choice in the space provided.**

_____ **10.** Hydrogen atoms may someday provide fuel for

**a.** nuclear fission. **b.** hydroelectric power.
**c.** geothermal energy. **d.** nuclear fusion.

_____ **11.** Most solar collectors require

**a.** coal. **b.** water. **c.** fission. **d.** wind.

_____ **12.** Energy experts have harnessed geothermal energy by

**a.** building dams. **b.** building wind generators.
**c.** drilling wells. **d.** burning coal.

_____ **13.** In a hydroelectric plant, running water produces energy by spinning a

**a.** turbine. **b.** fan. **c.** windmill. **d.** reactor.

## Critical Thinking

**Read each question or statement and answer it in the space provided.**

**1.** What might a geologist from a mining company look for in rock masses to identify possible copper deposits? Explain your answer.

______________________________________________

______________________________________________

______________________________________________

______________________________________________

______________________________________________

**2.** Do you think it would be profitable for a mining company to mine hematite? Explain your answer.

______________________________________________

______________________________________________

______________________________________________

______________________________________________

______________________________________________

## Chapter 11

**Read each question or statement and answer it in the space provided.**

**3.** A certain area has extensive deposits of shale. Why might a petroleum geologist be interested in the area?

**4.** If the United States continues using petroleum in vast amounts, it will become even more dependent on foreign sources for this resource. Why?

**5.** A certain company in your area produces U-235 pellets and fuel rods. With which energy source is the company involved? How do you know?

## Chapter 11

## Application

**Read each question or statement and answer it in the space provided.**

**1.** Imagine that you are on a committee to reduce air pollution in a crowded city. Would you recommend the use of high-sulfur coal as a fuel? Why or why not?

______________________________________________

______________________________________________

______________________________________________

______________________________________________

______________________________________________

**2.** Imagine that your senator is thinking of proposing a bill to cut off funding for nuclear fusion research. What might you say in a letter to change the senator's mind?

______________________________________________

______________________________________________

______________________________________________

______________________________________________

______________________________________________

**3.** If your school were converting to solar energy, how would you decide the best location on the roof to place a solar collector?

______________________________________________

______________________________________________

______________________________________________

______________________________________________

______________________________________________

**4.** If you were to construct a **concept map** using the new terms listed on page 208 of your textbook, what other word would you use as the top entry?

______________________________________________

Chapter 12

# Weathering and Erosion

## Review

**Choose the best response. Write the letter of that choice in the space provided.**

_____ **1.** Mechanical weathering of exposed surfaces causes rock to

**a.** decompose. **b.** break into smaller pieces.
**c.** melt. **d.** become buried.

_____ **2.** A common kind of mechanical weathering is called

**a.** oxidation. **b.** carbonation. **c.** ice wedging. **d.** leaching.

_____ **3.** Oxides of sulfur and nitrogen that combine with water vapor cause

**a.** iron rain. **b.** acid rain. **c.** mechanical weathering. **d.** carbonation.

_____ **4.** The mineral in igneous rock that is least affected by chemical weathering is

**a.** calcite. **b.** quartz. **c.** feldspar. **d.** mica.

_____ **5.** The surface area of rocks exposed to weathering is increased by

**a.** burial. **b.** leaching. **c.** quartz. **d.** joints.

_____ **6.** Chemical weathering is most rapid in

**a.** hot, dry climates. **b.** cold, dry climates.
**c.** cold, wet climates. **d.** hot, wet climates.

_____ **7.** The chemical composition of soil depends to a large extent on

**a.** topography. **b.** its A horizon. **c.** the parent material. **d.** its B horizon.

_____ **8.** The soil in tropical climates is often

**a.** thick. **b.** dry. **c.** thin. **d.** fertile.

_____ **9.** The transport of weathered materials by a moving natural agent is called

**a.** mass movement. **b.** weathering.
**c.** erosion. **d.** creep.

_____ **10.** All of the following farming methods prevent gullying, except

**a.** terracing. **b.** contour plowing.
**c.** strip-cropping. **d.** irrigation.

## Chapter 12

**Choose the best response. Write the letter of that choice in the space provided.**

_____ **11.** The most effective of all mass movements is

**a.** a landslide. **b.** a rockfall. **c.** a mudflow. **d.** creep.

_____ **12.** When a mountain reaches old age, it is eroded to an almost featureless surface called a

**a.** peneplain. **b.** monadnock. **c.** talus slope. **d.** mesa.

## Critical Thinking

**Read each question or statement and answer it in the space provided.**

**1.** Compare the weathering processes that affect a rock on top of a mountain and a rock buried beneath the ground.

______________________________________________

______________________________________________

______________________________________________

______________________________________________

**2.** Which do you think would weather faster, a sculptured marble statue or a smooth marble column? Explain your answer.

______________________________________________

______________________________________________

______________________________________________

______________________________________________

**3.** Compare the appearance of a 50-year-old highway that runs through a desert with that of a highway of the same age that runs through New York City.

______________________________________________

______________________________________________

______________________________________________

______________________________________________

## Chapter 12

**Read each question or statement and answer it in the space provided.**

**4.** If the A horizon and B horizon of a soil are relatively red, what do you think the underlying C horizon is composed of? Why?

______________________________________________

______________________________________________

______________________________________________

______________________________________________

**5.** Mudflows in the southern California hills are usually preceded by a dry summer and widespread fires, followed by torrential rainfall. Explain why.

______________________________________________

______________________________________________

______________________________________________

______________________________________________

**6.** How can you determine whether a plateau in an advanced stage of evolution is made of sandstone? Explain.

______________________________________________

______________________________________________

______________________________________________

______________________________________________

**7.** Suppose that a mountain has been wearing down at the rate of about 2 cm per year for 10 years. In the eleventh year, scientists find that the mountain is no longer losing elevation. What do you think has happened?

______________________________________________

______________________________________________

______________________________________________

______________________________________________

## Chapter 12

## Application

**Read each question or statement and answer it in the space provided.**

**1.** You are consulted on whether a limestone building or a quartzite building would better withstand the effects of acid rain. What would be your response and why?

______________________________________________________________________

______________________________________________________________________

______________________________________________________________________

______________________________________________________________________

______________________________________________________________________

______________________________________________________________________

**2.** If you were a farmer and could choose the ideal climate in which to grow your crops of deep-rooted plants, what climate would you choose? Explain your reasons.

______________________________________________________________________

______________________________________________________________________

______________________________________________________________________

______________________________________________________________________

______________________________________________________________________

______________________________________________________________________

**3.** Suppose you wanted to cultivate grapevines on a hillside in Italy. What farming methods would you use? Explain why.

______________________________________________________________________

______________________________________________________________________

______________________________________________________________________

______________________________________________________________________

______________________________________________________________________

______________________________________________________________________

# Chapter 13
# Water and Erosion

## Review

**Choose the best response. Write the letter of that choice in the space provided.**

_____ **1.** The change of water vapor into liquid water is called

**a.** runoff. **b.** evaporation. **c.** desalination. **d.** condensation.

_____ **2.** Vegetation gives off water vapor into the atmosphere through a process called

**a.** condensation. **b.** rejuvenation. **c.** saltation. **d.** transpiration.

_____ **3.** In a water budget, the income is precipitation and the expense is

**a.** evapotranspiration and runoff. **b.** condensation and saltation.
**c.** erosion and conservation. **d.** rejuvenation and sedimentation.

_____ **4.** The process that turns sea water into fresh water is

**a.** desalination. **b.** transpiration. **c.** conservation. **d.** rejuvenation.

_____ **5.** The land areas from which water runs off into a stream are called its

**a.** tributaries. **b.** divides. **c.** watersheds. **d.** gullies.

_____ **6.** Tributaries branch out and lengthen as a river system develops by

**a.** headward erosion. **b.** condensation.
**c.** saltation. **d.** runoff.

_____ **7.** The stream load that includes gravel and large rocks is the

**a.** suspended load. **b.** dissolved load. **c.** runoff load. **d.** bed load.

_____ **8.** When a young river deepens its channel faster than it can cut into its sides, the result is

**a.** a gradient. **b.** a V-shaped valley.
**c.** a floodway. **d.** an oxbow lake.

_____ **9.** A stream whose gradient has been increased by movement of the earth's crust is said to be

**a.** rejuvenated. **b.** meandering. **c.** eroded. **d.** suspended.

_____ **10.** The triangular formation that occurs when a stream deposits its sediment at the base of a steep slope is called

**a.** a delta. **b.** a meander. **c.** an oxbow lake. **d.** an alluvial fan.

Name ______________________ Class ______________ Date ______________

**MODERN EARTH SCIENCE**

## Chapter 13

**Choose the best response. Write the letter of that choice in the space provided.**

_____ **11.** The part of a valley floor that may be covered during a flood becomes the

**a.** floodway. **b.** groundwater. **c.** floodplain. **d.** artificial levee.

_____ **12.** One indirect method of flood control is

**a.** soil conservation.
**b.** dams.
**c.** floodways.
**d.** artificial levees.

## Critical Thinking

**Read each question or statement and answer it in the space provided.**

**1.** How would the earth's water cycle be affected if a significant portion of the sun's rays were blocked by dust or other contaminants in the atmosphere?

______________________________________________

______________________________________________

______________________________________________

______________________________________________

**2.** How might the local water budget of Calcutta differ from that of Stockholm? Use an atlas to determine the geographic location of these two cities before thinking about your answer.

______________________________________________

______________________________________________

______________________________________________

______________________________________________

**3.** Desalination may someday provide an almost endless supply of fresh water. What other problem must be solved before the desalinated water can be widely used?

______________________________________________

______________________________________________

______________________________________________

______________________________________________

## Chapter 13

**Read each question or statement and answer it in the space provided.**

**4.** In the desert areas of the southwestern United States, there are many shallow, narrow ditches that cut through the landscape. What do you suppose these ditches are? What was the most likely cause of their formation? If these ditches were located anywhere else, what might happen to them? Why does this not happen in the desert?

______________________________________________

______________________________________________

______________________________________________

______________________________________________

______________________________________________

______________________________________________

**5.** The Colorado River is usually grayish-brown as it flows through the Grand Canyon. What causes this color?

______________________________________________

______________________________________________

______________________________________________

______________________________________________

______________________________________________

______________________________________________

**6.** Why do you think the surface of an alluvial fan is sloping and that of a delta is flat?

______________________________________________

______________________________________________

______________________________________________

______________________________________________

______________________________________________

______________________________________________

## Chapter 13

## Application

**Read each question or statement and answer it in the space provided.**

**1.** Assume that you decided to test the color of the water of the Colorado River over a period of years and found that it was becoming clearer. What would you conclude was happening to the river?

______________________________

______________________________

______________________________

______________________________

**2.** If you were trying to locate a mature river for geologic exploration, what characteristics would you look for?

______________________________

______________________________

______________________________

______________________________

**3.** Developers are planning to build retirement communities on the floodplain of a river, but away from the banks. Considering only the safety aspect, would you argue for or against building these communities? Support your argument.

______________________________

______________________________

______________________________

______________________________

**4.** What steps might the developers in Question 3 take to protect the people and property in the communities?

______________________________

______________________________

______________________________

______________________________

Chapter 14

# Groundwater and Erosion

## Review

**Choose the best response. Write the letter of that choice in the space provided.**

_____ **1.** Any body of rock through which water can flow and which can store enough water for domestic or industrial use is called

**a.** a well. **b.** an aquifer.
**c.** a sinkhole. **d.** an artesian formation.

_____ **2.** The percentage of open space in a given volume of rock is called its

**a.** viscosity. **b.** permeability. **c.** capillary fringe. **d.** porosity

_____ **3.** When all the particles in a sediment are about the same size, the sediment is said to be

**a.** fractured. **b.** well sorted. **c.** permeable. **d.** poorly sorted.

_____ **4.** The ease with which water can pass through a rock or sediment is called

**a.** permeability. **b.** porosity. **c.** carbonation. **d.** velocity.

_____ **5.** The underground layer of rock where all of the open spaces are filled with water is called the

**a.** zone of aeration. **b.** cap rock.
**c.** water table. **d.** zone of saturation.

_____ **6.** The upper surface of the zone of saturation is called the

**a.** cap rock. **b.** water table. **c.** gradient. **d.** travertine.

_____ **7.** The slope of a water table is its

**a.** gradient. **b.** permeability. **c.** porosity. **d.** aquifer.

_____ **8.** It takes a long time for polluted groundwater to become pure again because

**a.** groundwater can only be replaced artificially.
**b.** groundwater is replenished slowly.
**c.** groundwater travels very quickly.
**d.** groundwater can never be replaced.

_____ **9.** A natural flow of groundwater that has reached the surface is

**a.** a spring. **b.** an aquifer. **c.** a well. **d.** a travertine.

**Chapter 14**

**Choose the best response. Write the letter of that choice in the space provided.**

_____ **10.** Pumping water from a well causes a local lowering of the water table known as a

**a.** cone of depression.
**b.** horizontal fissure.
**c.** hot spring.
**d.** sinkhole.

_____ **11.** Hot springs that form in areas of recent volcanic activity may become

**a.** caverns. **b.** natural bridges. **c.** artesian formations. **d.** mud pots.

_____ **12.** Travertine usually forms

**a.** terraces. **b.** natural bridges. **c.** sinkholes. **d.** caverns.

_____ **13.** When part of the roof of a cavern collapses, the result is a

**a.** sinkhole. **b.** stalactite. **c.** horizontal fissure. **d**. geyser.

_____ **14.** Calcite formations suspended from the ceiling of a cavern are called

**a.** stalagmites. **b.** stalactites. **c.** sinks. **d.** aquifers.

_____ **15.** Regions where the results of chemical weathering by groundwater are clearly visible are said to have

**a.** sink topography.
**b.** karst topography.
**c.** limestone topography.
**d.** artesian formations.

## Critical Thinking

**Read each question or statement and answer it in the space provided.**

**1.** A rock can be porous, yet impermeable. Explain how.

_______________________________________________

_______________________________________________

_______________________________________________

_______________________________________________

_______________________________________________

_______________________________________________

_______________________________________________

_______________________________________________

Name ______________________ Class ______________ Date ______________

MODERN EARTH SCIENCE

## Chapter 14

**Read each question or statement and answer it in the space provided.**

**2.** In areas where the water table is at the surface of the land, what type of terrain would you expect to find?

______________________________________________

______________________________________________

______________________________________________

______________________________________________

**3.** Describe an artesian formation and explain how the water in an artesian well may have entered the ground many hundreds of kilometers away.

______________________________________________

______________________________________________

______________________________________________

______________________________________________

**4.** Explain how a mud pot forms. Why are mud pots sometimes called paint pots?

______________________________________________

______________________________________________

______________________________________________

______________________________________________

**5.** Explain the process that results in the formation of stalactites and stalagmites. Can you think of another process in nature that produces shapes similar to stalactites?

______________________________________________

______________________________________________

______________________________________________

______________________________________________

## Chapter 14

**Read each question or statement and answer it in the space provided.**

**6.** Do you think an area with karst topography would have many or few surface streams? Explain your answer.

______________________________________________________________

______________________________________________________________

______________________________________________________________

## Application

**Read each question or statement and answer it in the space provided.**

**1.** You are having an unusually dry summer. What effect would you expect this dry season to have on the capillary action in the soil?

______________________________________________________________

______________________________________________________________

______________________________________________________________

**2.** Your city has been experiencing shortages in its water supply. What would be your advice on replenishing the groundwater artificially? What are your reasons for making this suggestion?

______________________________________________________________

______________________________________________________________

______________________________________________________________

**3.** Construct a **concept map** using 10 of the new terms listed on page 272 of your textbook by making connections that illustrate the relationship among the terms.

# Chapter 15

# Glaciers and Erosion

## Review

**Choose the best response. Write the letter of that choice in the space provided.**

_____ **1.** Snow accumulates year after year in polar regions and

**a.** in southern Canada. **b.** in the United States.
**c.** below the snowline. **d.** at high elevations.

_____ **2.** Glaciers formed in mountainous areas are called

**a.** continental ice sheets. **b.** valley glaciers.
**c.** icebergs. **d.** ice shelves.

_____ **3.** Antarctica is covered by the earth's largest

**a.** iceberg. **b.** valley glacier.
**c.** continental ice sheet. **d.** outlet glacier.

_____ **4.** A glacier will move by sliding when the base of the ice and rock are separated by a thin layer of

**a.** water. **b.** snow. **c.** pebbles. **d.** drift.

_____ **5.** When a glacier moves by internal plastic flow,

**a.** its center moves fastest.
**b.** its bottom moves fastest.
**c.** its edges move fastest.
**d.** the whole ice mass moves at the same speed.

_____ **6.** Glacial erosion may produce a bowl-shaped depression known as

**a.** a moraine. **b.** an esker. **c.** a cirque. **d.** a horn.

_____ **7.** As a glacier moves through a valley, it carves out

**a.** a U-shape. **b.** a V-shape. **c.** an esker. **d.** a moraine.

_____ **8.** Unsorted glacial deposits are called

**a.** stratified drift. **b.** outwash plains. **c.** eskers. **d.** till.

_____ **9.** Long, winding ridges of gravel and sand deposited by a meltwater stream under the ice are called

**a.** eskers. **b.** drumlins. **c.** outwash plains. **d.** medial moraines.

## Chapter 15

**Choose the best response. Write the letter of that choice in the space provided.**

_____ **10.** A kettle is a

**a.** hill. **b.** depression. **c.** ridge. **d.** mound.

_____ **11.** A deposit of stratified drift is called

**a.** a drumlin. **b.** outwash.
**c.** a ground moraine. **d.** a roche moutonnée.

_____ **12.** Many glacial lakes are formed in

**a.** outwash plains. **b.** kettles. **c.** eskers. **d.** arêtes.

_____ **13.** During the last glacial period, the average temperature was about

**a.** 5°C lower than today. **b.** 15°C lower than today.
**c.** 35°C lower than today. **d.** 50°C lower than today.

_____ **14.** One component of the Milankovitch theory is

**a.** the circular motion of the earth's axis. **b.** continental drift.
**c.** volcanic activity. **d.** landslide activity.

_____ **15.** A proposed cause of the ice ages is decreased solar energy reaching the earth due to

**a.** a lunar eclipse. **b.** blockage by volcanic dust.
**c.** sinking of the land. **d.** increased storm activity.

## Critical Thinking

**Read each question or statement and answer it in the space provided.**

**1.** Imagine that there is a village in Alaska located at the edge of a glaciated mountain range. During the year there is an unusually large amount of snowfall. How might this snowfall affect the valley glaciers? What danger might this pose for the inhabitants of the village?

______________________________________________

______________________________________________

______________________________________________

______________________________________________

______________________________________________

______________________________________________

## Chapter 15

**Read each question or statement and answer it in the space provided.**

**2.** Why is it important for scientists to monitor and study the continental ice sheets that cover Greenland and Antarctica?

______________________________________________________________________

______________________________________________________________________

______________________________________________________________________

______________________________________________________________________

______________________________________________________________________

______________________________________________________________________

**3.** Antarctic explorers need special training to travel safely over the ice sheet. Besides the cold, what structural aspects of the glaciers might be dangerous?

______________________________________________________________________

______________________________________________________________________

______________________________________________________________________

______________________________________________________________________

______________________________________________________________________

______________________________________________________________________

**4.** In addition to decreasing temperature and increasing snowfall, what other phenomenon might signal an impending ice age?

______________________________________________________________________

______________________________________________________________________

______________________________________________________________________

______________________________________________________________________

______________________________________________________________________

______________________________________________________________________

Name ______________________ Class ____________ Date ____________



# Chapter 15

## Application

**Read each question or statement and answer it in the space provided.**

**1.** List some features caused by glacial erosion and deposition that you might see on a car trip.

______________________________________________________________________

______________________________________________________________________

______________________________________________________________________

______________________________________________________________________

______________________________________________________________________

______________________________________________________________________

**2.** In what ways might past glacial action in New England and New York State affect tourism and recreation in those regions today?

______________________________________________________________________

______________________________________________________________________

______________________________________________________________________

______________________________________________________________________

______________________________________________________________________

______________________________________________________________________

**3.** A group of scientists is trying to find evidence to support Milankovitch's theory of the earth's ice ages. List some kinds of research they might need to do.

______________________________________________________________________

______________________________________________________________________

______________________________________________________________________

______________________________________________________________________

______________________________________________________________________

______________________________________________________________________

Name ______________________ Class ______________ Date ______________



Chapter 16

# Erosion by Wind and Waves

## Review

**Choose the best response. Write the letter of that choice in the space provided.**

_____ **1.** Loose fragments of rock and minerals measuring between 0.06 mm and 2 mm are referred to as

**a.** pollen. **b.** dust. **c.** sand. **d.** desert pavement.

_____ **2.** Wind moves sand by

**a.** saltation. **b.** abrasion. **c.** emergence. **d.** depression.

_____ **3.** The most common form of wind erosion is

**a.** migration. **b.** abrasion. **c.** saltation. **d.** deflation.

_____ **4.** Dunes move primarily by

**a.** abrasion. **b.** deflation. **c.** migration. **d.** submergence.

_____ **5.** Unlayered, yellowish, fine-grained deposits are called

**a.** beaches. **b.** loess. **c.** dunes. **d.** desert pavement.

_____ **6.** The most important erosion agent along shorelines is

**a.** waves. **b.** wind. **c.** weathering. **d.** tides.

_____ **7.** All of the following shoreline features are produced by wave erosion of sea cliffs except

**a.** spits. **b.** sea stacks.
**c.** wave-cut terraces. **d.** sea arches.

_____ **8.** The composition of beach deposits depends on

**a.** the climate. **b.** the source rock.
**c.** the time of year. **d.** wave action.

_____ **9.** Longshore-current deposition of sand at the end of a headland produces a

**a.** sand bar. **b.** dune. **c.** spit. **d.** sea cliff.

_____ **10.** Sea level is now

**a.** stationary. **b.** falling about 1 mm/yr.
**c.** rising about 1 cm/yr. **d.** rising about 1 mm/yr.

## Chapter 16

**Choose the best response. Write the letter of that choice in the space provided.**

_____ **11.** A coastline resulting from a rise in sea level or subsidence of the land is

**a.** emergent. **b.** submergent. **c.** glaciated. **d.** volcanic.

_____ **12.** Barrier islands tend to migrate

**a.** seaward. **b.** along the shore. **c.** in the summer. **d.** toward the shore.

_____ **13.** Coral reefs are produced by marine organisms that

**a.** swim in circles.
**b.** extract minerals from sea water to form hard skeletons.
**c.** attach themselves to sand bars.
**d.** build nests of sand.

_____ **14.** If shoreline resources continue to be used as in the past,

**a.** coastal land will be greatly improved. **b.** the damage can be repaired.
**c.** sea level will fall. **d.** coastal land will become less desirable.

## Critical Thinking

**Read each question or statement and answer it in the space provided.**

**1.** The deserts of the southwestern United States contain many tall, sculpted rock formations that are the result of weathering and erosion. Was wind or water erosion the more likely agent responsible for these formations? Explain your answer.

______________________________

______________________________

______________________________

______________________________

**2.** Beautifully colored sunsets and sunrises are the result of dust in the atmosphere. Such sunsets and sunrises were visible around the world for two years after Krakatau, a volcanic island in Indonesia, erupted in 1883. Explain how this is possible.

______________________________

______________________________

______________________________

______________________________

## Chapter 16

**Read each question or statement and answer it in the space provided.**

**3.** Supose that once each month for a year a satellite orbiting the earth takes a photograph of the same 1-$km^2$ sandy area of the Sahara. Would the surface features shown in these 12 photographs remain essentially the same, or would they be very different? Explain your answer.

**4.** What effect does the development of a wave-built terrace have on erosion of the shoreline? What phenomenon might counteract this effect?

**5.** Describe two ways that the melting of continental glaciers affects the relative level of the land and the sea.

## Chapter 16

## Application

**Read each question or statement and answer it in the space provided.**

**1.** Suppose that scientists have observed that the size of the sand bars in a particular area has been decreasing steadily over the past 20 years. What does this observation suggest about the climate of the area? Explain why.

______________________________________________

______________________________________________

______________________________________________

______________________________________________

**2.** As you approach a large landmass by ship, you notice an island connected to the shore by a tombolo. What type of shoreline do you predict the mainland will have? Explain your answer.

______________________________________________

______________________________________________

______________________________________________

______________________________________________

**3.** Construct a **concept map** using the new terms listed on page 312 of your textbook to show the relationship between landforms and the agents of erosion that produce them.

Name ____________________ Class ____________ Date ____________



Chapter 17

# The Rock Record

## Review

**Choose the best response. Write the letter of that choice in the space provided.**

_____ **1.** The concept that the present is the key to the past is part of the

**a.** type of unconformity.
**b.** law of superposition.
**c.** law of crosscutting relationships.
**d.** principle of uniformitarianism.

_____ **2.** A sedimentary rock layer is older than the layers above it and younger than the layers below it, according to the

**a.** type of unconformity.
**b.** law of superposition.
**c.** law of crosscutting relationships.
**d.** principle of uniformitarianism.

_____ **3.** A gap in the sequence of rock layers is

**a.** a bedding plane.
**b.** a varve.
**c.** an unconformity.
**d.** a uniformity.

_____ **4.** An unconformity that results when new sediments are deposited on eroded horizontal layers is

**a.** an angular unconformity.
**b.** a disconformity.
**c.** crosscut unconformity.
**d.** a nonconformity.

_____ **5.** A fault or intrusion is younger than the rock it cuts through, according to the

**a.** type of unconformity.
**b.** law of superposition.
**c.** law of crosscutting relationships.
**d.** principle of uniformitarianism.

_____ **6.** The age of a rock in years is known as

**a.** index age.
**b.** relative age.
**c.** half-life age.
**d.** absolute age.

_____ **7.** Varves are formed by layers of

**a.** limestone mixed with coarse sediments.
**b.** coarse sediments followed by fine sediments.
**c.** shale followed by sandstone.
**d.** sandstone followed by shale.

_____ **8.** An atom with a mass number of 234 and an atomic number of 90 has

**a.** 90 neutrons and 144 protons.
**b.** 234 neutrons and 90 protons.
**c.** 144 neutrons and 90 protons.
**d.** 117 neutrons and 117 protons.

Name ______________________ Class ______________ Date ______________



## Chapter 17

**Choose the best response. Write the letter of that choice in the space provided.**

_____ **9.** The process whereby the remains of an organism are preserved by drying is

**a.** petrification. **b.** mummification. **c.** erosion. **d.** superposition.

_____ **10.** Molds filled with sediments sometimes produce

**a.** casts. **b.** gastroliths. **c.** coprolites. **d.** imprints.

_____ **11.** Fossils that are found in many parts of the earth and that were formed by organisms that lived during a brief period of geologic time are known as

**a.** coprolites. **b.** molds. **c.** gastroliths. **d.** index fossils.

## Critical Thinking

**Read each question or statement and answer it in the space provided.**

**1.** Hutton developed the principle of uniformitarianism by observing geologic changes on his farm. What changes might he have observed?

______________________________________________

______________________________________________

______________________________________________

______________________________________________

______________________________________________

______________________________________________

**2.** How might a scientist determine the original positions of the sedimentary layers beneath an angular unconformity?

______________________________________________

______________________________________________

______________________________________________

______________________________________________

______________________________________________

______________________________________________

## Chapter 17

**Read each question or statement and answer it in the space provided.**

**3.** Of two intrusions, one cuts through all the rock layers. The other is eroded and lies beneath several layers of sedimentary rock. Which intrusion is younger? Why?

______________________________________________

______________________________________________

______________________________________________

______________________________________________

______________________________________________

______________________________________________

**4.** What information about past climatic changes might scientists gather by studying varves?

______________________________________________

______________________________________________

______________________________________________

______________________________________________

______________________________________________

______________________________________________

**5.** How are the processes of mummification and freezing similar?

______________________________________________

______________________________________________

______________________________________________

______________________________________________

______________________________________________

______________________________________________

## Chapter 17

## Application

**Read each question or statement and answer it in the space provided.**

**1.** A scientist is attempting to calculate the absolute age of sedimentary rock layers by using the average rate of deposition. However, between the third and fourth layers is a disconformity. What difficulties might this cause in determining the absolute ages of the layers below the disconformity?

______________________________________________

______________________________________________

______________________________________________

______________________________________________

______________________________________________

**2.** How many years would it take for 16 g of U-238 to decay into 0.5 g of U-238 and 15.5 g of daughter products?

______________________________________________

______________________________________________

______________________________________________

______________________________________________

______________________________________________

**3.** Suppose that a certain type of fossil with unusual features is found in many areas of the earth. It represents a brief span of geologic time but occurs only in small numbers. Would the fossil make a good index fossil? Explain.

______________________________________________

______________________________________________

______________________________________________

______________________________________________

______________________________________________

## Chapter 18

# A View of the Earth's Past

### Review

**Choose the best response. Write the letter of that choice in the space provided.**

_____ **1.** The geologic time scale is a

**a.** scale for weighing rocks.
**b.** calendar used by geologists.
**c.** rock record of the earth's past.
**d.** collection of the same kind of rocks.

_____ **2.** Scientists have been able to determine the absolute ages of most rock layers in the geologic column by using

**a.** the law of superposition.
**b.** radiometric dating.
**c.** rates of deposition.
**d.** rates of erosion.

_____ **3.** An event that geologists would use in dividing the geologic time scale into smaller units is

**a.** the eruption of a volcano.
**b.** a change in rock color.
**c.** a change in sediment type.
**d.** the arrival of an ice age.

_____ **4.** To determine the age of a specific rock, scientists might correlate it with a layer in the geologic column that has the same relative position, physical characteristics, and

**a.** fossil content. **b.** weight. **c.** temperature. **d.** density.

_____ **5.** Paleozoic means

**a.** ancient life. **b.** middle life. **c.** recent life. **d.** primitive life.

_____ **6.** Geologic periods may be divided into

**a.** eras. **b.** epochs. **c.** days. **d.** months.

_____ **7.** Cenozoic means

**a.** ancient life. **b.** middle life. **c.** recent life. **d.** primitive life.

_____ **8.** Precambrian time ended about

**a.** 4.6 billion years ago.
**b.** 540 million years ago.
**c.** 65 million years ago.
**d.** 25 thousand years ago.

_____ **9.** The most common fossils found in Precambrian rocks are

**a.** graptolites. **b.** trilobites. **c.** eurypterids. **d.** stromatolites.

## Chapter 18

**Choose the best response. Write the letter of that choice in the space provided.**

_____ **10.** An important invertebrate in the Paleozoic Era was the

**a.** ammonite. **b.** trilobite. **c.** ostracoderm. **d.** stromatolite.

_____ **11.** The first vertebrates appeared during

**a.** Precambrian time.
**b.** the Paleozoic Era.
**c.** the Mesozoic Era.
**d.** the Cenozoic Era.

_____ **12.** The Age of Fishes is the name commonly given to the

**a.** Cambrian Period.
**b.** Ordovician Period.
**c.** Silurian Period.
**d.** Devonian Period.

_____ **13.** The Age of Reptiles is the name commonly given to

**a.** Precambrian time.
**b.** the Paleozoic Era.
**c.** the Mesozoic Era.
**d.** the Cenozoic Era.

_____ **14.** Dinosaurs became the dominant life-form during the

**a.** Permian Period.
**b.** Silurian Period.
**c.** Jurassic Period.
**d.** Tertiary Period.

_____ **15.** The first flowering plants made their appearance during the

**a.** Cretaceous Period.
**b.** Triassic Period.
**c.** Carboniferous Period.
**d.** Ordovician Period.

_____ **16.** The Age of Mammals is the name commonly given to

**a.** Precambrian time.
**b.** the Paleozoic Era.
**c.** the Mesozoic Era.
**d.** the Cenozoic Era.

## Critical Thinking

**Read each question or statement and answer it in the space provided.**

**1.** Explain how the law of superposition has aided scientists in the development of the geologic column.

______________________________________________

______________________________________________

______________________________________________

______________________________________________

Name ______________________ Class ______________ Date ______________

**MODERN EARTH SCIENCE**

## Chapter 18

**Read each question or statement and answer it in the space provided.**

2. Why is it difficult to divide Precambrian time into periods?

3. Paleontologists classify fossils by similarities in the structure of the hard parts of the organisms. Sometimes, however, scientists find that this is an unreliable method of classifying fossils. Explain.

4. There are rich deposits of coal in the Appalachian Mountains. During which geologic era were these deposits probably formed? Explain your answer.

5. What information in the geologic record might lead scientists to infer that shallow seas covered much of the earth during the Paleozoic Era?

## Chapter 18

**Read each question or statement and answer it in the space provided.**

6. Explain the basis scientists may have used for dividing the Cenozoic Era into the Tertiary and Quaternary periods.

## Application

**Read each question or statement and answer it in the space provided.**

1. In 1966, two amateur fossil hunters found a fossil on the shore at Cliffwood, New Jersey. Scientists determined that the fossil was an insect intermediate between primitive ants and wasps. What information might scientists gain from this fossil discovery?

2. If you found a rock layer that contained dinosaur fossils, what assumptions could you make about the age of the rock layer?

3. Referring to the geologic time scale, construct a **concept map** that illustrates the relationship between the various time divisions and some of the organisms that first developed during those divisions.

Chapter 19

# The History of the Continents

## Review

**Choose the best response. Write the letter of that choice in the space provided.**

_____ **1.** Areas of exposed Precambrian rocks that may represent ancient continents are called

**a.** shields. **b.** fossils. **c.** outcrops. **d.** cratons.

_____ **2.** During the Paleozoic Era, the continents were joined together in a huge landmass called

**a.** Panthalassa. **b.** Laurasia. **c.** Pangaea. **d.** Gondwanaland.

_____ **3.** North America and Eurasia were once part of a landmass called

**a.** Panthalassa. **b.** Laurasia. **c.** Tethys. **d.** Gondwanaland.

_____ **4.** South America, Africa, India, Australia, and Antarctica were once part of a landmass called

**a.** Panthalassa. **b.** Laurasia. **c.** Tethys. **d.** Gondwanaland.

_____ **5.** Pangaea began to break up about

**a.** 650 million years ago. **b.** 450 million years ago.
**c.** 250 million years ago. **d.** 150 million years ago.

_____ **6.** The area of exposed Precambrian rocks found in North America is called the

**a.** New England Shield. **b.** Canadian Shield.
**c.** American Shield. **d.** Appalachian Shield.

_____ **7.** During the Paleozoic Era, much of what is now the United States was covered by

**a.** a shallow sea. **b.** a desert. **c.** an ice sheet. **d.** grasslands.

_____ **8.** During the Mesozoic Era, new crust was added to western North America in the form of

**a.** shields. **b.** terranes. **c.** cratons. **d.** geosynclines.

_____ **9.** North America had taken on its present shape by the beginning of

**a.** Precambrian time. **b.** the Paleozoic Era.
**c.** the Mesozoic Era. **d.** the Cenozoic Era.

## Chapter 19

**Choose the best response. Write the letter of that choice in the space provided.**

_____ **10.** The Colorado Plateau was uplifted during

**a.** Precambrian time. **b.** the Paleozoic Era.
**c.** the Mesozoic Era. **d.** the Cenozoic Era.

_____ **11.** The absence of certain layers in the rock record of the Grand Canyon may be the result of

**a.** flooding. **b.** cross-bedding. **c.** erosion. **d.** deposition.

_____ **12.** The cross-bedded sandstone layers in the Grand Canyon indicate that the area may once have been covered by a

**a.** swamp. **b.** desert. **c.** glacier. **d.** forest.

_____ **13.** The layers of limestone and shale in the Grand Canyon indicate that the area was once covered by a

**a.** shallow sea. **b.** desert. **c.** glacier. **d.** forest.

_____ **14.** Fossils of marine organisms are often found in limestone and

**a.** sandstone. **b.** granite. **c.** lava. **d.** shale.

## Critical Thinking

**Read each question or statement and answer it in the space provided.**

**1.** One hundred fifty million years from now, the continents will all have drifted to new locations. How might these changes affect life on the earth?

## Chapter 19

**Read each question or statement and answer it in the space provided.**

**2.** Assume scientists know the rate at which North America and Eurasia are drifting farther apart on their respective tectonic plates. How might they determine when North America separated from Eurasia during the breakup of Pangaea?

______________________________________________

______________________________________________

______________________________________________

______________________________________________

______________________________________________

______________________________________________

**3.** If scientists discovered huge salt deposits in central Canada, what might they assume about the geologic history of that region?

______________________________________________

______________________________________________

______________________________________________

______________________________________________

______________________________________________

______________________________________________

**4.** If you found a bed that contains trilobites above a bed that contains mammal fossils, what might you assume about the beds? How could you find out for certain?

______________________________________________

______________________________________________

______________________________________________

______________________________________________

______________________________________________

______________________________________________

## Chapter 19

**Read each question or statement and answer it in the space provided.**

**5.** Geologists have found no deposits from the Pennsylvanian Period in the Grand Canyon. What might they assume about that period of the region's history?

______________________________________________

______________________________________________

______________________________________________

______________________________________________

______________________________________________

## Application

**Read each question or statement and answer it in the space provided.**

**1.** The Grand Canyon in Arizona and Zion and Bryce canyons in Utah are all located on the Colorado Plateau. The rocks at the base of the Grand Canyon are the oldest; the rocks at the base of Bryce Canyon are the youngest. Bryce Canyon is located farthest north. What can you assume about how these canyons formed?

______________________________________________

______________________________________________

______________________________________________

______________________________________________

______________________________________________

**2.** During an expedition to the Grand Canyon, you find traces of *Homo sapiens* dating to the Pleistocene Epoch. In which layer of the canyon, if any, would these traces most likely have been found? Explain your answer.

______________________________________________

______________________________________________

______________________________________________

______________________________________________

______________________________________________

Chapter 20

# The Ocean Basins

## Review

**Choose the best response. Write the letter of that choice in the space provided.**

_____ **1.** The largest of the oceans is the

**a.** Atlantic. **b.** Indian. **c.** Pacific. **d.** Arctic.

_____ **2.** The continental and oceanic crust that lies beneath the ocean waters makes up the

**a.** ocean floor. **b.** continental margin.
**c.** continental slope. **d.** abyssal plain.

_____ **3.** A self-propelled, free-moving submarine that is equipped for ocean research is a

**a.** turbidity. **b.** bathyscaph. **c.** bathysphere. **d.** guyot.

_____ **4.** A system that is used for determining the depth of the ocean floor is

**a.** guyot. **b.** bathysphere. **c.** radiolaria. **d.** sonar.

_____ **5.** Those portions of the ocean floor made up of continental crust are called

**a.** continental margins. **b.** abyssal plains.
**c.** mid-ocean ridges. **d.** trenches.

_____ **6.** Some submarine canyons were probably formed by

**a.** subduction zones. **b.** turbidity currents.
**c.** changes in sea level. **d.** the accumulation of sediments.

_____ **7.** The accumulation of sediments at the base of the continental slope is called the

**a.** trench. **b.** turbidity current.
**c.** continental margin. **d.** continental rise.

_____ **8.** The deepest parts of the ocean are called

**a.** trenches. **b.** submarine canyons.
**c.** abyssal plains. **d.** continental rises.

_____ **9.** Thick sediment layers rarely form on abyssal plains that are bordered by

**a.** trenches. **b.** continental margins.
**c.** guyots. **d.** submarine canyons.

Name ______________________ Class ____________ Date ____________



## Chapter 20

**Choose the best response. Write the letter of that choice in the space provided.**

_____ **10.** Volcanic mountains scattered along the ocean basin are

**a.** diatoms. **b.** seamounts. **c.** mid-ocean ridges. **d.** foraminifera.

_____ **11.** Submerged flat-topped seamounts are

**a.** guyots. **b.** diatoms. **c.** submersibles. **d.** bathyspheres.

_____ **12.** Large quantities of the inorganic sediment that makes up the continental rise come from

**a.** turbidity currents. **b.** earthquakes.
**c.** diatoms. **d.** nodules.

_____ **13.** Much of the silica on the ocean floor comes from

**a.** nodules. **b.** radiolaria and diatoms.
**c.** guyots. **d.** foraminifera.

_____ **14.** Potato-shaped lumps of minerals on the ocean floor are called

**a.** guyots. **b.** foraminifera. **c.** nodules. **d.** diatoms.

_____ **15.** Very fine particles of silt and clay that have settled to the ocean floor are called

**a.** muds. **b.** guyots. **c.** seamounts. **d.** nodules.

## Critical Thinking

**Read each question or statement and answer it in the space provided.**

**1.** Suppose you are studying photographs of a newly discovered planet that has several large bodies of water. The three largest bodies of water—A, B, and C—are connected and occupy 70 percent of the surface of the planet. Two other bodies of water—D and E—are not quite so large and are partly surrounded by land. A sixth body of water—F—is completely surrounded by land. Which of these bodies of water resemble the earth's oceans? Which resemble seas? What type of body of water on earth does F most closely resemble?

________________________________________

________________________________________

________________________________________

________________________________________

________________________________________

________________________________________

## Chapter 20

**Read each question or statement and answer it in the space provided.**

2. The exploration of the ocean depths has been compared with the exploration of outer space. What similarities exist between these two environments and the attempts by people to explore them?

3. How might the continental shelves be affected if another ice age were to occur?

4. What might be the eventual fate of seamounts as they are carried along on the spreading oceanic crust?

5. A certain fish is known to exist only in one particular river in the central United States. Explain how the fossilized remains of this fish might become part of the sediments on the ocean floor.

6. Explain how it is possible that some red clays on the ocean floor have been found to contain material from outer space.

Name ______________________ Class ______________ Date ______________

**MODERN EARTH SCIENCE**

## Chapter 20

## Application

**Read each question or statement and answer it in the space provided.**

**1.** Suppose you were searching off the east coast of the United States for the wreckage of an old Spanish galleon. Explain how sonar could aid your search.

______________________________________________

______________________________________________

______________________________________________

**2.** Lava erupts from a volcano in the Cascade Mountains along the west coast of the United States. One component of the lava are traces of minerals similar to those found in nodules on the floor of the Pacific Ocean. How could these minerals have become part of the lava?

______________________________________________

______________________________________________

______________________________________________

**3.** Construct a **concept map** starting with the term *ocean floor*. Add the new terms listed on page 402 of your textbook and make connections to illustrate the relationships among them.

Name ______________________ Class ______________ Date ______________

**MODERN EARTH SCIENCE**

Chapter 21

# Ocean Water

## Review

**Choose the best response. Write the letter of that choice in the space provided.**

_____ **1.** The gas that dissolves most easily in ocean water is

**a.** carbon dioxide. **b.** argon. **c.** nitrogen. **d.** oxygen.

_____ **2.** The amount of dissolved solids in ocean water is called its

**a.** salinity. **b.** nekton. **c.** plankton. **d.** density.

_____ **3.** A thin floating layer of ice that covers an ocean surface is called

**a.** an iceberg. **b.** a thermocline. **c.** pack ice. **d.** benthos.

_____ **4.** A zone of rapid water temperature change with increasing depth is called

**a.** a benthos. **b.** an abyssal zone. **c.** a bathyal zone. **d.** a thermocline.

_____ **5.** When liquid water is warmed, its density

**a.** increases. **b.** decreases. **c.** remains the same. **d.** doubles.

_____ **6.** Although most of the various wavelengths of visible light are absorbed by ocean water, the one wavelength that tends to be reflected is the color

**a.** violet. **b.** green. **c.** yellow. **d.** blue.

_____ **7.** The process by which surface water is blown farther offshore and nutrient-rich deep water rises to take its place is called

**a.** desalination. **b.** distillation. **c.** upwelling. **d.** aquaculture.

_____ **8.** Drifting marine plants and animals are known as

**a.** plankton. **b.** benthos. **c.** nekton. **d.** tube worms.

_____ **9.** Marine animals that can swim to search for food and avoid predators are called

**a.** phytoplankton. **b.** zooplankton. **c.** nekton. **d.** benthos.

_____ **10.** Sea anemones are an example of

**a.** zooplankton. **b.** nekton. **c.** benthos. **d.** phytoplankton.

## Chapter 21

**Choose the best response. Write the letter of that choice in the space provided.**

_____ **11.** Perhaps the most unstable ocean environment is the

**a.** intertidal zone. **b.** abyssal zone. **c.** bathyal zone. **d.** neritic zone.

_____ **12.** The pelagic zone that has abundant sunlight, a fairly constant temperature, and relatively low pressure is the

**a.** bathyal zone. **b.** neritic zone. **c.** abyssal zone. **d.** intertidal zone.

_____ **13.** Which of the following is not a method for producing fresh water by desalinating ocean water?

**a.** distillation **b.** evaporation **c.** reverse osmosis **d.** aquaculture

_____ **14.** Lumps of minerals on the ocean floor are called

**a.** nekton. **b.** nodules. **c.** benthos. **d.** plankton.

_____ **15.** Aquaculture is another name for

**a.** desalination. **b.** distillation.
**c.** ocean farming. **d.** rapid temperature changes.

_____ **16.** In the future, scientists may be able to fertilize the ocean through

**a.** distillation. **b.** artificial upwelling.
**c.** mining nodules. **d.** dissolving gases.

## Critical Thinking

**Read each question or statement and answer it in the space provided.**

**1.** Suppose that climatic conditions over the earth's oceans changed dramatically. The changes resulted in a complete absence of upwelling and wave action. Explain what would happen to the marine life and why.

________________________________________________

________________________________________________

________________________________________________

________________________________________________

________________________________________________

________________________________________________

Name ______________________ Class ______________ Date ______________

MODERN EARTH SCIENCE

## Chapter 21

**Read each question or statement and answer it in the space provided.**

**2.** How would a significant decrease in sunlight affect phytoplankton?

______________________________________________________________

______________________________________________________________

______________________________________________________________

______________________________________________________________

______________________________________________________________

______________________________________________________________

**3.** What impact would the decrease in sunlight discussed in Question 2 have on other forms of marine life?

______________________________________________________________

______________________________________________________________

______________________________________________________________

______________________________________________________________

______________________________________________________________

______________________________________________________________

**4.** When oceanographers first explored the deep ocean basin along mid-ocean ridges, they discovered a variety of marine life, including sightless crabs. Explain why these crabs are not handicapped by their sightlessness.

______________________________________________________________

______________________________________________________________

______________________________________________________________

______________________________________________________________

______________________________________________________________

______________________________________________________________

## Chapter 21

## Application

**Read each question or statement and answer it in the space provided.**

**1.** Suppose you are contracted to build a desalination system for the state of Florida. In order to make your system as efficient as possible, you want to locate it in an area of the ocean where the salinity of the water is low. Where along the coast of Florida would you choose to build the site, near Miami or Tallahassee?

______________________________________________

______________________________________________

______________________________________________

______________________________________________

______________________________________________

**2.** You have collected samples of ocean water from the three locations: a site just off the coast of Israel; a site on Prudhoe Bay, off the coast of northern Alaska; and a site off the coast of Sweden, on the Baltic Sea. All three samples have been brought to room temperature. From which location would the water have the greatest density? Why?

______________________________________________

______________________________________________

______________________________________________

______________________________________________

______________________________________________

**3.** You have decided to start an aquatic farm. In which of the zones of ocean life should you locate your farm? Why?

______________________________________________

______________________________________________

______________________________________________

______________________________________________

______________________________________________

Name ____________ Class ____________ Date ____________



Chapter 22

# Movements of the Ocean

## Review

**Choose the best response. Write the letter of that choice in the space provided.**

_____ **1.** The waters in the ocean move in giant streams called

**a.** currents. **b.** westerlies. **c.** waves. **d.** tides.

_____ **2.** The effect of the earth's rotation on winds and ocean currents is called the

**a.** neap-tide effect. **b.** refraction effect.
**c.** Coriolis effect. **d.** tsunami effect.

_____ **3.** Two warm-water currents, the North and South Equatorial currents, flow

**a.** northward. **b.** westward.
**c.** southward. **d.** eastward.

_____ **4.** The warm, swift current that flows up the east coast of the United States and then into the North Atlantic is the

**a.** Japan Current. **b.** California Current.
**c.** Labrador Current. **d.** Gulf Stream.

_____ **5.** A weak current is known as a

**a.** swell. **b.** drift. **c.** tide. **d.** crest.

_____ **6.** The vast area of relatively still, warm water located in the middle of the North Atlantic is referred to as the

**a.** Sargasso Sea. **b.** North Atlantic Drift.
**c.** North Equatorial Current. **d.** Gulf Stream.

_____ **7.** A strong, turbid current that is caused by an underwater landslide is called

**a.** a surface current. **b.** an equatorial current.
**c.** a turbidity current. **d.** a tsunami.

_____ **8.** Deep currents are the result of

**a.** the Coriolis effect. **b.** changes in the density of ocean water.
**c.** the trade winds. **d.** neap tides.

_____ **9.** The periodic up-and-down movement of water is a

**a.** current. **b.** fetch. **c.** breaker. **d.** wave.

## Chapter 22

**Choose the best response. Write the letter of that choice in the space provided.**

_____ **10.** The highest point of a wave is its

**a.** trough. **b.** crest. **c.** period. **d.** length.

_____ **11.** The time it takes for one complete wavelength to pass a given point is called the

**a.** wave speed. **b.** wave height. **c.** trough. **d.** wave period.

_____ **12.** The length of open water across which the wind blows is the

**a.** trough. **b.** sargassum. **c.** fetch. **d.** wave period.

_____ **13.** When the faster-moving top part of a wave topples over the slower bottom part, the result is a foamy mass of water called a

**a.** whitecap. **b.** rip current. **c.** breaker. **d.** crest.

_____ **14.** Sand bars may be formed by

**a.** undertows. **b.** rip currents. **c.** tsunamis. **d.** longshore currents.

_____ **15.** The daily changes in the elevation of the ocean surface are called

**a.** waves. **b.** tides. **c.** swells. **d.** breakers.

_____ **16.** The difference in the level between high tide and low tide is the

**a.** neap tide. **b.** spring tide. **c.** tidal range. **d.** tidal oscillation.

_____ **17.** The movement of water toward and away from the coasts due to tidal forces is called a

**a.** tidal bore. **b.** tidal current. **c.** tidal range. **d.** tidal oscillation.

## Critical Thinking

**Read each question or statement and answer it in the space provided.**

1. During winter in the northern Indian Ocean, winds called *monsoons* blow in a direction opposite to the direction that they blow during summer. What effect do these winds have on the surface currents?

______________________________________________________________

______________________________________________________________

______________________________________________________________

______________________________________________________________

## Chapter 22

**Choose the best response. Write the letter of that choice in the space provided.**

**2.** Suppose that a retaining wall is built along a shoreline. What will happen to waves as they pass over the retaining wall?

______________________________________________

______________________________________________

______________________________________________

______________________________________________

______________________________________________

______________________________________________

**3.** Imagine that you are fishing from a small boat anchored off the shore of the Gulf of Mexico. You are lulled to sleep by the gently rocking boat but wake up to find your boat on wet sand. What happened?

______________________________________________

______________________________________________

______________________________________________

______________________________________________

______________________________________________

______________________________________________

**4.** Along the shoreline of a particular bay, high tide rises to about 20 m and low tide recedes to about 10 m. The bay is wide, with a large opening into the ocean. Should a tidal-power plant be built across the bay? Explain your answer.

______________________________________________

______________________________________________

______________________________________________

______________________________________________

______________________________________________

______________________________________________

## Chapter 22

## Application

**Read each question or statement and answer it in the space provided.**

**1.** What could you do to create a deep current in a pan of slightly salty water? Explain how you would go about this.

______________________________________________________________________

______________________________________________________________________

______________________________________________________________________

______________________________________________________________________

______________________________________________________________________

______________________________________________________________________

______________________________________________________________________

______________________________________________________________________

______________________________________________________________________

**2.** From the deck of a stationary ship on the open ocean you notice that at least 15 minutes pass between the rise of one wave and the next. Can you assume that the sea is very calm? Explain why or why not.

______________________________________________________________________

______________________________________________________________________

______________________________________________________________________

______________________________________________________________________

______________________________________________________________________

______________________________________________________________________

______________________________________________________________________

______________________________________________________________________

______________________________________________________________________

Name ______________________ Class ______________ Date ______________

MODERN EARTH SCIENCE

Chapter 23

# The Atmosphere

## Review

**Choose the best response. Write the letter of that choice in the space provided.**

_____ **1.** During one part of the nitrogen cycle, nitrogen is removed from the air mainly by nitrogen-fixing

**a.** bacteria. **b.** waves. **c.** minerals. **d.** crystals.

_____ **2.** Standard atmospheric pressure measured at sea level is

**a.** 99 percent. **b.** 1.03 kg. **c.** 32 km. **d.** 101,325 $N/m^2$.

_____ **3.** A barometer measures

**a.** atmospheric pressure. **b.** wind speed.
**c.** ozone concentration. **d.** wavelengths.

_____ **4.** Almost all of the water and carbon dioxide in the atmosphere is in the

**a.** exosphere. **b.** ionosphere. **c.** troposphere. **d.** stratopause.

_____ **5.** Radio stations can increase the distances they reach by bouncing radio waves off the

**a.** stratosphere. **b.** tropopause. **c.** ionosphere. **d.** troposphere.

_____ **6.** Around Los Angeles, frequent temperature inversions are the result of cool, polluted air being trapped by

**a.** acid rain. **b.** a layer of warmer air.
**c.** mountains. **d.** the ocean.

_____ **7.** Almost all of the energy reaching the earth from the sun is in the form of

**a.** atomic particles. **b.** electromagnetic waves.
**c.** ultraviolet rays. **d.** gamma rays.

_____ **8.** Raindrops may separate sunlight into a range of colors, thereby causing

**a.** a mirage. **b.** an inferior mirage.
**c.** acid precipitation. **d.** a rainbow.

_____ **9.** The process in which the atmosphere traps warming solar rays at the earth's surface is called the

**a.** greenhouse effect. **b.** Coriolis effect.
**c.** doldrums. **d.** convection cell.

## Chapter 23

**Choose the best response. Write the letter of that choice in the space provided.**

_____ **10.** Heat can be transferred within the atmosphere in three ways—radiation, conduction, and

**a.** scattering.
**b.** temperature inversion.
**c.** weathering.
**d.** convection.

_____ **11.** A vertical looping pattern of air flow is known as

**a.** the Coriolis effect.
**b.** a convection cell.
**c.** a trade wind.
**d.** a westerly.

_____ **12.** A gentle wind covering less than 100 km is called

**a.** a jet stream. **b.** the doldrums. **c.** a breeze. **d.** a trade wind.

## Critical Thinking

**Read each question or statement and answer it in the space provided.**

**1.** Explain how houseplants can increase the amount of oxygen in your home.

**2.** During a jet flight over the North Pole and toward a region in the middle latitudes, the pilot adjusts the altimeter. Why is this adjustment necessary?

## Chapter 23

**Read each question or statement and answer it in the space provided.**

**3.** Most aerosol sprays are banned in the United States. Which of the four layers of the atmosphere does this ban help protect? Explain your answer.

______________________________________________

______________________________________________

______________________________________________

______________________________________________

**4.** You hear a lecture about the earth's weather. The speaker says, "Infrared rays coming from the earth's surface heat the atmosphere much like a greenhouse is heated." Explain why that statement is incorrect.

______________________________________________

______________________________________________

______________________________________________

______________________________________________

**5.** What effect might jet streams have on airplane travel?

______________________________________________

______________________________________________

______________________________________________

______________________________________________

**6.** If there is a breeze blowing from the ocean to the land on the coast of Maine, about what time of day is it? How do you know?

______________________________________________

______________________________________________

______________________________________________

______________________________________________

## Chapter 23
## Application

**Read each question or statement and answer it in the space provided.**

**1.** In a certain area of the country, many of the fish in a local lake have died. In addition, soils were found to be highly acidic, and nearby trees were losing their leaves. What kind of pollution may have caused these problems? What is the source of this pollution?

______________________________________________

______________________________________________

______________________________________________

______________________________________________

______________________________________________

**2.** In a drive across the desert with your family, you see a distant sand dune that appears to be floating on air. How can you explain what you see?

______________________________________________

______________________________________________

______________________________________________

______________________________________________

______________________________________________

**3.** In what ways would a knowledge of the global wind belts have helped a sixteenth-century explorer sailing between Spain and the northern part of South America?

______________________________________________

______________________________________________

______________________________________________

______________________________________________

______________________________________________

Chapter 24

# Water in the Atmosphere

## Review

**Choose the best response. Write the letter of that choice in the space provided.**

_____ **1.** The process by which ice changes directly into water vapor is referred to as

**a.** advection. **b.** conduction. **c.** sublimation. **d.** condensation.

_____ **2.** Relative humidity is always expressed

**a.** in g/m$^3$. **b.** as a percentage. **c.** in g/kg. **d.** in degrees Celsius.

_____ **3.** To express the actual amount of moisture in the air, meteorologists use

**a.** latent heat. **b.** relative humidity.
**c.** specific humidity. **d.** dew point.

_____ **4.** When air temperature drops, its capacity for holding water is

**a.** slightly higher. **b.** much higher. **c.** about the same. **d.** lower.

_____ **5.** The tiny water droplets that result when air is cooled by contact with a cold surface are called

**a.** dew. **b.** frost. **c.** humidity. **d.** steam fog.

_____ **6.** Changes in temperature that result solely from the expansion or compression of air are called

**a.** adiabatic. **b.** supercooling. **c.** advection. **d.** latent cooling.

_____ **7.** Clouds form when the water vapor in air condenses as

**a.** the air is heated. **b.** the air is cooled.
**c.** snow falls. **d.** the air is superheated.

_____ **8.** Low, sheetlike clouds are called

**a.** cirrus clouds. **b.** stratus clouds.
**c.** cumulus clouds. **d.** cirrocumulus clouds.

_____ **9.** The term *nimbo-* or *-nimbus* added to the name of any form of cloud means

**a.** high. **b.** billowy. **c.** rain. **d.** layered.

## Chapter 24

**Choose the best response. Write the letter of that choice in the space provided.**

_____ **10.** The fog that results from the nightly cooling of the earth is called

**a.** steam fog. **b.** upslope fog. **c.** radiation fog. **d.** advection fog.

_____ **11.** Rain that freezes when it strikes the ground produces

**a.** sleet. **b.** glaze ice. **c.** hail. **d.** frost.

_____ **12.** Clouds in which the water droplets remain liquid below 0°C are said to be

**a.** saturated. **b.** supersaturated. **c.** superheated. **d.** supercooled.

_____ **13.** In one method of cloud seeding, silver-iodide crystals are used as

**a.** freezing nuclei. **b.** dry ice. **c.** cloud droplets. **d.** a superheater.

_____ **14.** A wide-mouthed funnel and a cylindrical container are used in making an instrument called a

**a.** hygrometer. **b.** rain gauge. **c.** barometer. **d.** psychrometer.

## Critical Thinking

**Read each question or statement and answer it in the space provided.**

**1.** Where would the air contain the most moisture—over Panama or over Antarctica? Explain your answer.

______________________________________________

______________________________________________

______________________________________________

______________________________________________

**2.** One body of air has a relative humidity of 97 percent. Another has a relative humidity of 44 percent. At the same temperature, which body of air is closer to its dew point? Explain your answer.

______________________________________________

______________________________________________

______________________________________________

______________________________________________

## Chapter 24

**Read each question or statement and answer it in the space provided.**

**3.** Why would polluted air more likely form clouds than clean air would?

______________________________________________

______________________________________________

______________________________________________

______________________________________________

______________________________________________

______________________________________________

**4.** In the tropical regions, raindrops are commonly formed by coalescence. Little precipitation there forms by supercooling. Why might this be true?

______________________________________________

______________________________________________

______________________________________________

______________________________________________

______________________________________________

______________________________________________

**5.** One day in January, 6 cm of very wet snow falls on your area. If all this snow melted quickly, how deep would the water be? Explain how you know.

______________________________________________

______________________________________________

______________________________________________

______________________________________________

______________________________________________

______________________________________________

## Chapter 24

## Application

**Read each question or statement and answer it in the space provided.**

**1.** You are in school and you hear thunder outside. Describe the clouds that you would probably see if you looked out the classroom window.

______________________________________________________________________

______________________________________________________________________

______________________________________________________________________

______________________________________________________________________

______________________________________________________________________

______________________________________________________________________

**2.** You are camping in a valley on a calm, clear night. You awake and notice a thick fog. What type of fog is it, and how did it form?

______________________________________________________________________

______________________________________________________________________

______________________________________________________________________

______________________________________________________________________

______________________________________________________________________

______________________________________________________________________

**3.** You see lumps of ice the size of golf balls falling from the sky. Are you seeing sleet or hail? How do you know?

______________________________________________________________________

______________________________________________________________________

______________________________________________________________________

______________________________________________________________________

______________________________________________________________________

______________________________________________________________________

# Chapter 25

# Weather

## Review

**Choose the best response. Write the letter of that choice in the space provided.**

_____ **1.** A region where air masses can form must be fairly

**a.** cold. **b.** warm. **c.** hilly. **d.** uniform.

_____ **2.** In an air mass designated *cP*, the *c* stands for

**a.** continental. **b.** cold. **c.** coastal. **d.** cool.

_____ **3.** Continental polar Canadian air masses generally move

**a.** southeasterly. **b.** northerly. **c.** northeasterly. **d.** westerly.

_____ **4.** The air masses that sometimes bring heavy rains to the deserts of the southwestern United States are called

**a.** continental polar Canadian. **b.** maritime polar Atlantic.
**c.** maritime tropical Pacific. **d.** maritime continental tropical.

_____ **5.** The type of front formed when two air masses move parallel to the front between them is called

**a.** stationary. **b.** occluded. **c.** polar. **d.** warm.

_____ **6.** A front that is completely lifted off the ground by cold air is called

**a.** cold. **b.** occluded. **c.** polar. **d.** warm.

_____ **7.** The winds of a wave cyclone blow in circular paths around a

**a.** front. **b.** low-pressure center.
**c.** high-pressure center. **d.** jet stream.

_____ **8.** The eye of a hurricane is a region of

**a.** hailstorms. **b.** torrential rainfall.
**c.** calm, clear air. **d.** strong winds.

_____ **9.** In the mature stage of a thunderstorm, a cumulus cloud grows until it becomes a

**a.** stratocumulus cloud. **b.** altocumulus cloud.
**c.** cumulonimbus cloud. **d.** cirrocumulus cloud.

## Chapter 25

**Choose the best response. Write the letter of that choice in the space provided.**

_____ **10.** Tornadoes that occur over the ocean are called

**a.** waterspouts. **b.** typhoons. **c.** waves. **d.** hurricanes.

_____ **11.** A wind with a direction designated as 90° is blowing from the

**a.** north. **b.** south. **c.** east. **d.** west.

_____ **12.** An instrument package attached to a weather balloon is

**a.** an anemometer. **b.** a wind vane. **c.** a radiosonde. **d.** a thermograph.

_____ **13.** The lines on a weather map connecting points of equal atmospheric pressure are called

**a.** isobars. **b.** isotherms. **c.** highs. **d.** lows.

_____ **14.** It is generally difficult to accurately predict

**a.** wind speed. **b.** amount of precipitation.
**c.** wind direction. **d.** temperature.

_____ **15.** Of the following terms, which one would most likely be placed higher on a concept map designed for this chapter?

**a.** occluded front **b.** wave cyclone **c.** tornado **d.** air mass

## Critical Thinking

**Read each question or statement and answer it in the space provided.**

**1.** If the air in your region is warm and dry, what type of air mass could be responsible? What letters designate this air mass?

_______________________________________________

_______________________________________________

_______________________________________________

_______________________________________________

_______________________________________________

_______________________________________________

_______________________________________________

Name ____________________ Class ____________ Date ____________



## Chapter 25

**Read each question or statement and answer it in the space provided.**

**2.** People on Vancouver Island, off the west coast of Canada, hear reports of a wave cyclone in the Gulf of Alaska. Is it likely that the wave cyclone will reach their area? Explain why.

______________________________________________

______________________________________________

______________________________________________

______________________________________________

**3.** Suppose a hurricane is passing over a Caribbean island. Suddenly, the rain and winds stop and the air becomes calm and clear. Is it safe to go outside? Explain your answer.

______________________________________________

______________________________________________

______________________________________________

______________________________________________

**4.** Is it safe to be on the street in an automobile during a tornado? Explain your answer.

______________________________________________

______________________________________________

______________________________________________

______________________________________________

**5.** In what direction would a wind of 315° make a wind vane point?

______________________________________________

______________________________________________

______________________________________________

______________________________________________

## Chapter 25

**Read each question or statement and answer it in the space provided.**

**6.** An air traffic controller is monitoring nearby airplanes by radar. The controller warns an incoming pilot of a storm a few miles away. How did radar help the controller detect the storm?

______________________________________________

______________________________________________

______________________________________________

______________________________________________

## Application

**Read each question or statement and answer it in the space provided.**

**1.** Suppose you are traveling with friends through the desert in the southwestern United States and a thunderstorm occurs. You then tell them about the type of air mass that may have brought the storm. What did you say?

______________________________________________

______________________________________________

______________________________________________

**2.** Suppose the air is warm and moist. You hear on the weather report, however, that a fast-moving cold front will reach your region the next day. What kind of weather conditions can you expect?

______________________________________________

______________________________________________

______________________________________________

**3.** On a weather map, you see a station model with the circle half darkened and a straight vertical line extending upward. What can you say about the weather in that area for that day?

______________________________________________

______________________________________________

______________________________________________

Chapter 26

# Climate

## Review

**Choose the best response. Write the letter of that choice in the space provided.**

_____ **1.** At the equator the sun's rays always strike the earth

**a.** at a low angle.
**b.** at nearly a 90° angle.
**c.** 18 hours each day.
**d.** no more than 8 hours each day.

_____ **2.** Nights are longest in the winter and shortest in the summer

**a.** at the equator.
**b.** at high altitudes.
**c.** in the middle of the ocean.
**d.** at the poles.

_____ **3.** Water cools

**a.** more slowly than land does.
**b.** more quickly than land does.
**c.** through the process of transpiration.
**d.** because of waves and currents.

_____ **4.** Ocean currents influence temperature by

**a.** eroding shorelines.
**b.** heating or cooling the air.
**c.** washing warm, dry sediments out to sea.
**d.** dispersing the rays of the sun.

_____ **5.** Winds that blow in opposite directions in different seasons because of the differential heating of the land and the oceans are called

**a.** chinooks. **b.** mistrals. **c.** monsoons. **d.** wave cyclones.

_____ **6.** When a moving air mass encounters a mountain range, it

**a.** stops moving. **b.** slows and sinks. **c.** rises and cools. **d.** reverses its direction.

_____ **7.** Tropical deserts exhibit all of the following characteristics except

**a.** location between 20° and 30° latitude.
**b.** dense plant growth.
**c.** influence of the subtropical highs.
**d.** extremely dry conditions.

_____ **8.** A tropical climate that is characterized by very wet summers and very dry winters is called

**a.** a Mediterranean climate.
**b.** a savanna climate.
**c.** a trade-wind climate.
**d.** an equatorial climate.

## Chapter 26

**Choose the best response. Write the letter of that choice in the space provided.**

_____ **9.** In regions with a Mediterranean climate, almost all the yearly precipitation falls

**a.** during monsoons.
**b.** in the summer.
**c.** in the winter.
**d.** during hurricanes.

_____ **10.** Weather conditions tend to fluctuate rapidly throughout the year in a

**a.** subarctic climate.
**b.** middle-latitude desert climate.
**c.** Mediterranean climate.
**d.** humid continental climate.

_____ **11.** The various structures and activities in cities affect the local climate by

**a.** decreasing the average temperature.
**b.** increasing both the average temperature and precipitation.
**c.** increasing the average temperature and decreasing the precipitation.
**d.** decreasing the precipitation.

## Critical Thinking

**Read each question or statement and answer it in the space provided.**

**1.** The Milankovitch theory states that a periodic change in the tilt of the earth's axis was one factor in the onset of the ice ages. Use what you know about the factors affecting climate to explain how an ice age might have occurred.

______________________________________________

______________________________________________

______________________________________________

______________________________________________

**2.** If you enjoy a warm, dry climate all year but do not like the sparse vegetation in deserts, in what climate or climates would you live? Assume that you would not mind moving one or more times a year.

______________________________________________

______________________________________________

______________________________________________

______________________________________________

## Chapter 26

**Read each question or statement and answer it in the space provided.**

**3.** Explain why the vegetation in areas with a tundra climate is sparse, even though these areas receive precipitation that is adequate to support plant life.

______________________________________________

______________________________________________

______________________________________________

______________________________________________

______________________________________________

______________________________________________

**4.** Why do weather conditions change rapidly in a humid continental climate and remain relatively constant in the other middle-latitude climates?

______________________________________________

______________________________________________

______________________________________________

______________________________________________

______________________________________________

______________________________________________

**5.** Explain why the classification of climates often fails when you think only in terms of a specific location.

______________________________________________

______________________________________________

______________________________________________

______________________________________________

______________________________________________

______________________________________________

## Chapter 26

## Application

**Read each question or statement and answer it in the space provided.**

**1.** Imagine you are going to build a vacation house near a coast with a warm offshore current. What must you investigate to determine how the current will affect the temperature on land? Explain your answer.

______________________________________________

______________________________________________

______________________________________________

______________________________________________

**2.** Suppose your family was moving to the mountains, but you do not like humid weather. Should you encourage them to find a house on the side of the mountains facing toward or away from the prevailing winds? Explain why.

______________________________________________

______________________________________________

______________________________________________

______________________________________________

**3.** Construct a **concept map** that correlates the various climates you learned about in this chapter with their geographical locations on the earth.

# Chapter 27

# Stars and Galaxies

## Review

**Choose the best response. Write the letter of that choice in the space provided.**

_____ **1.** In the majority of stars, the most common element is

**a.** oxygen. **b.** helium. **c.** hydrogen. **d.** sodium.

_____ **2.** The color of the hottest stars is

**a.** red. **b.** yellow. **c.** green. **d.** blue.

_____ **3.** Stars appear to move in circular paths around Polaris because

**a.** the earth rotates on its axis.
**b.** the earth orbits the sun.
**c.** the stars revolve around Polaris.
**d.** Polaris is the center of the Milky Way Galaxy.

_____ **4.** The change in position of a nearby star compared with the position of a faraway star is called

**a.** parallax. **b.** red shift. **c.** blue shift. **d.** a Cepheid variable.

_____ **5.** The brightest stars have apparent magnitudes that are

**a.** over +20. **b.** between +10 and +19.
**c.** between +1 and +9. **d.** negative numbers.

_____ **6.** The absolute magnitude of a star is

**a.** the relative brightness of the star.
**b.** the true brightness of the star.
**c.** the comparative brightness of the star.
**d.** the apparent brightness of the star.

_____ **7.** A protostar becomes a star when it begins to

**a.** develop a red shift. **b.** generate energy.
**c.** shrink and spin. **d.** explode as a nova.

_____ **8.** A main-sequence star generates energy by fusing

**a.** nitrogen into iron. **b.** helium into carbon.
**c.** hydrogen into helium. **d.** nitrogen into carbon.

Name ______________________ Class ______________ Date ______________

MODERN EARTH SCIENCE

## Chapter 27

**Choose the best response. Write the letter of that choice in the space provided.**

_____ **9.** A dying star can shed some of its gases as a

**a.** planetary nebula.
**b.** white dwarf.
**c.** globular cluster.
**d.** supernova.

_____ **10.** Black holes are difficult to locate because they

**a.** move very quickly.
**b.** do not give off light.
**c.** have very low gravity.
**d.** are far away from any stars.

_____ **11.** A pattern of stars is called a

**a.** galaxy. **b.** nebula. **c.** pulsar. **d.** constellation.

_____ **12.** Stars appear in fixed locations in the sky because they

**a.** are so far from the earth.
**b.** do not move.
**c.** are all moving toward the earth.
**d.** are all in the same galaxy.

_____ **13.** The basic types of galaxies are

**a.** spiral, elliptical, and irregular.
**b.** barred, elliptical, and open.
**c.** spiral, quasar, and pulsar.
**d.** open, binary, and globular.

_____ **14.** Quasar formation is associated with

**a.** nuclear fusion.
**b.** main-sequence stars.
**c.** the explosion of a supernova.
**d.** the early universe.

## Critical Thinking

**Read each question or statement and answer it in the space provided.**

**1.** If the spectrum of a star indicates that the star shines with a red light, approximately what is the surface temperature of the star?

______________________________________________________________________

______________________________________________________________________

______________________________________________________________________

______________________________________________________________________

______________________________________________________________________

______________________________________________________________________

Name ______________________ Class ______________ Date ______________



## Chapter 27

**Read each question or statement and answer it in the space provided.**

**2.** Why are different constellations visible during different seasons of the year?

______________________________________________

______________________________________________

______________________________________________

______________________________________________

**3.** Explain why Polaris is considered to be a very bright star even though it is not a bright star in the earth's sky.

______________________________________________

______________________________________________

______________________________________________

______________________________________________

**4.** Why does heat build up more rapidly in a massive protostar than in a less massive one?

______________________________________________

______________________________________________

______________________________________________

______________________________________________

**5.** Explain why an old main-sequence star will be composed of a higher percentage of helium than will a young main-sequence star.

______________________________________________

______________________________________________

______________________________________________

______________________________________________

## Chapter 27

**Read each question or statement and answer it in the space provided.**

**6.** If all galaxies began to show blue shifts, what would this indicate about the fate of the universe?

______________________________________________

______________________________________________

______________________________________________

______________________________________________

## Application

**Read each question or statement and answer it in the space provided.**

**1.** If you determined that a certain star displayed a large parallax, what could you say about its distance from the earth?

______________________________________________

______________________________________________

______________________________________________

______________________________________________

**2.** Suppose that a scientist has discovered a red-dwarf star. Describe the likely size and surface temperature of such a star.

______________________________________________

______________________________________________

______________________________________________

______________________________________________

**3.** When looking through a nearby university's telescope, you observed a galaxy that has no young stars and contains little dust or gas. What kind of galaxy were you probably looking at?

______________________________________________

______________________________________________

______________________________________________

______________________________________________

Chapter 28

# The Sun

## Review

**Choose the best response. Write the letter of that choice in the space provided.**

_____ **1.** According to Einstein's theory of relativity, in the formula $E = mc^2$, the $c$ stands for

**a.** corona. **b.** core. **c.** the speed of light. **d.** the length of time.

_____ **2.** A nuclear reaction in which two atomic nuclei combine is called

**a.** fission. **b.** fusion. **c.** magnetism. **d.** granulation.

_____ **3.** The portion of the sun in which energy moves from atom to atom in the form of waves is called the

**a.** radiative zone. **b.** convective zone. **c.** solar wind. **d.** chromosphere.

_____ **4.** The portion of the sun normally visible from the earth is the

**a.** core. **b.** photosphere. **c.** corona. **d.** solar nebula.

_____ **5.** The sunspot cycle lasts

**a.** 2 years. **b.** 5 years. **c.** 11 years. **d.** 19 years.

_____ **6.** Sudden outward eruptions of electrically charged atomic particles from the sun are called

**a.** planetesimals. **b.** coronas. **c.** sunspots. **d.** solar flares.

_____ **7.** Gusts of solar wind can cause

**a.** protoplanets. **b.** magnetic storms. **c.** nuclear fission. **d.** nuclear fusion.

_____ **8.** Northern lights and southern lights are other names for

**a.** prominences. **b.** auroras. **c.** granulations. **d.** protoplanets.

_____ **9.** The hypothesis that the sun and the planets developed out of the same cloud of gas and dust is called the

**a.** nebular theory. **b.** theory of relativity. **c.** nuclear theory. **d.** theory of convection.

_____ **10.** The small bodies of matter that filled the solar nebula are called

**a.** protoplanets. **b.** planetesimals. **c.** auroras. **d.** proton nuclei.

## Chapter 28

**Choose the best response. Write the letter of that choice in the space provided.**

_____ **11.** Compared with the size of the present-day planets, the protoplanets were

**a.** much smaller. **b.** slightly smaller. **c.** similar. **d.** larger.

_____ **12.** The first atmosphere of the earth had a high percentage of

**a.** helium. **b.** oxygen. **c.** nitrogen. **d.** water vapor.

_____ **13.** In the process of photosynthesis, cyanobacteria and green plants give off

**a.** oxygen. **b.** carbon dioxide. **c.** hydrogen. **d.** helium.

_____ **14.** Water vapor began to condense into oceans about

**a.** 1 million to 1.5 million years ago. **b.** 1 billion to 1.5 billion years ago.
**c.** 3 billion to 3.5 billion years ago. **d.** 15 billion to 15.5 billion years ago.

## Critical Thinking

**Read each question or statement and answer it in the space provided.**

**1.** Explain how the transfer of energy in a pan of hot water is similar to the transfer of energy in the sun's convective zone.

______________________________________________________________

______________________________________________________________

______________________________________________________________

______________________________________________________________

______________________________________________________________

**2.** Predict what would happen to the number of sunspots if the sun's magnetic field suddenly increased in strength.

______________________________________________________________

______________________________________________________________

______________________________________________________________

______________________________________________________________

______________________________________________________________

Name ____________ Class ____________ Date ____________

MODERN EARTH SCIENCE

## Chapter 28

**Read each question or statement and answer it in the space provided.**

**3.** If the earth's magnetosphere shifted, what would happen to the area where auroras are most often visible?

**4.** How would the layers of the earth be different if the planet had never been hotter than it is today?

**5.** How would the atmosphere of the earth be different if the earth had formed from a much larger protoplanet?

## Chapter 28

## Application

**Read each question or statement and answer it in the space provided.**

**1.** You are asked to explain the following statement to an elementary school class: Solar hydrogen is responsible for life existing on the earth. What would you say?

______________________________________________

______________________________________________

______________________________________________

______________________________________________

**2.** A group of senior citizens has decided to head north to view the northern lights. When would you advise them to go in order to see the most frequent displays?

______________________________________________

______________________________________________

______________________________________________

______________________________________________

**3.** Using the new terms listed on page 586 of your textbook, construct a **concept map** that, among other things, illustrates the path of energy from the core of the sun to the far side of the earth.

Name ______________________ Class ____________ Date ____________

MODERN EARTH SCIENCE

# Chapter 29

# The Solar System

## Review

**Choose the best response. Write the letter of that choice in the space provided.**

_____ **1.** Ptolemy modified Aristotle's model of the universe to include

**a.** Oort clouds. **b.** retrograde motion. **c.** comets. **d.** shooting stars.

_____ **2.** Copernicus's model of the solar system differed from Ptolemy's because it was

**a.** geocentric. **b.** lunocentric. **c.** ethnocentric. **d.** heliocentric.

_____ **3.** Kepler's first law states that each planet orbits the sun in a path called

**a.** an ellipse. **b.** a circle. **c.** an epicycle. **d.** a period.

_____ **4.** Kepler's law that describes how fast planets travel at different points in their orbits is called the law of

**a.** ellipses. **b.** equal speed. **c.** equal areas. **d.** periods.

_____ **5.** The weak magnetic field around Mercury suggests

**a.** volcanic activity. **b.** a dense atmosphere. **c.** a core of molten iron. **d.** that it is located close to the sun.

_____ **6.** The planet that rotates in a direction that is opposite the direction of the other planets is

**a.** Mercury. **b.** Venus. **c.** the earth. **d.** Mars.

_____ **7.** The tilt of the axis of Mars is nearly the same as that of

**a.** Mercury. **b.** Venus. **c.** the earth. **d.** Jupiter.

_____ **8.** The planet that rotates faster than any other planet in the solar system is

**a.** the earth. **b.** Jupiter. **c.** Uranus. **d.** Pluto.

_____ **9.** The most distinctive feature of Jupiter is its

**a.** Great Red Spot. **b.** Great Dark Spot. **c.** rings. **d.** elongated orbit.

## Chapter 29

**Choose the best response. Write the letter of that choice in the space provided.**

_____ **10.** All of the outer planets in the solar system are large except

**a.** Saturn. **b.** Uranus. **c.** Neptune. **d.** Pluto.

_____ **11.** The asteroid belt exists in a region between the orbits of

**a.** Mercury and Venus. **b.** Venus and the earth.
**c.** the earth and Mars. **d.** Mars and Jupiter.

_____ **12.** The composition of asteroids suggests that they are

**a.** small moons.
**b.** fragments of planetesimals.
**c.** the nuclei of comets.
**d.** environments that possibly can support life.

_____ **13.** Meteoroids can provide information about

**a.** the composition of the solar nebula before the earth and its moon formed.
**b.** the size of the earth.
**c.** the destiny of the solar system.
**d.** the size of the universe.

## Critical Thinking

**Read each question or statement and answer it in the space provided.**

**1.** Assume that an intelligent life-form exists on Pluto—the planet with the longest orbit period in the solar system. Would astronomers on Pluto be likely to propose a heliocentric model of the solar system? Explain your answer.

______________________________________________

______________________________________________

______________________________________________

______________________________________________

______________________________________________

## Chapter 29

**Read each question or statement and answer it in the space provided.**

**2.** If you know the distance from the sun to a planet, what other information can you determine about the orbit of the planet? Explain your answer.

**3.** The surfaces of some asteroids reflect only small amounts of light. Other asteroids reflect up to 40 percent of the light falling on them. Of what materials would each type of asteroid probably be composed?

**4.** By constructing a **concept map,** you make connections that illustrate relationships among certain terms. How would doing so assist your understanding of this chapter?

## Chapter 29

## Application

**Read each question or statement and answer it in the space provided.**

**1.** Suppose that a new planet has just been discovered. It has no rings or moons and has a surface pitted with impact craters. In what group of planets do you think this planet is located? Explain how you know.

**2.** What type of core do you predict that the new planet mentioned in Question 1 will have?

**3.** Suppose you live in an unglaciated area and have found a chunk of rock that you suspect might be a stony meteoroid. What data would help you verify your hypothesis?

Name ________________ Class ________________ Date ________________



Chapter 30

# Moons and Rings

## Review

**Choose the best response. Write the letter of that choice in the space provided.**

_____ **1.** Dark areas on the moon that are smooth and reflect little light are called

**a.** rilles. **b.** maria. **c.** rays. **d.** breccia.

_____ **2.** Most of the information astronomers have gathered about the interior of the moon has come from

**a.** telescopes. **b.** satellites. **c.** spectrographs. **d.** seismographs.

_____ **3.** Soon after the moon formed, it was covered with

**a.** water. **b.** anorthosites. **c.** frozen hydrogen. **d.** molten rock.

_____ **4.** In the most recent stage in the development of the moon,

**a.** the densest material sank to the core.
**b.** the crust began to break.
**c.** the earth's gravity captured the moon.
**d.** the number of meteoroids hitting the moon decreased.

_____ **5.** The moon is closest to the earth at

**a.** new moon. **b.** full moon. **c.** perigee. **d.** apogee.

_____ **6.** During each orbit around the earth, the moon spins on its axis

**a.** 1 time. **b.** about 27 times. **c.** about 29 times. **d.** 365 times.

_____ **7.** In a lunar eclipse, the moon

**a.** casts a shadow on the earth. **b.** is in the earth's shadow.
**c.** is between the earth and the sun. **d.** blocks part of the sun from view.

_____ **8.** When the size of the visible portion of the moon is decreasing, the moon is

**a.** full. **b.** annular. **c.** waxing. **d.** waning.

_____ **9.** In the crescent phases, the entire moon shines dimly because of

**a.** light produced by the earth.
**b.** sunlight reflected off the earth.
**c.** hydrogen fusion in the core of the moon.
**d.** energy produced by the rotation of the moon.

## Chapter 30

**Choose the best response. Write the letter of that choice in the space provided.**

_____ **10.** The two moons of Mars are

**a.** Io and Europa.
**b.** Titan and Charon.
**c.** Phobos and Deimos.
**d.** Triton and Nereid.

_____ **11.** Compared with the other moons of Jupiter, the four Galilean moons are

**a.** larger.
**b.** farther from Jupiter.
**c.** denser.
**d.** younger.

_____ **12.** The rings of Saturn are probably composed of

**a.** regolith.
**b.** small pieces of black rock.
**c.** several hundred small moons.
**d.** billions of pieces of ice and rock.

## Critical Thinking

**Read each question or statement and answer it in the space provided.**

**1.** How would the craters on the moon be different today if the moon had developed an atmosphere that had wind and contained water?

**2.** If meteoroids had stopped hitting the moon before the outer surface of the moon cooled, why would the maria not have developed?

## Chapter 30

**Read each question or statement and answer it in the space provided.**

**3.** Suppose that the moon spun twice on its axis during each orbit around the earth. How would study of the moon from the earth be easier?

______________________________________________

______________________________________________

______________________________________________

______________________________________________

______________________________________________

______________________________________________

**4.** Venus does not cause a solar eclipse even though it passes between the earth and the sun. Instead, it appears as a black dot moving across the face of sun. Explain why this happens.

______________________________________________

______________________________________________

______________________________________________

______________________________________________

______________________________________________

______________________________________________

**5.** Would a satellite orbiting the earth go through phases like those of the moon? Explain your answer.

______________________________________________

______________________________________________

______________________________________________

______________________________________________

______________________________________________

______________________________________________

## Chapter 30

## Application

**Read each question or statement and answer it in the space provided.**

**1.** Your friend who just got a new telescope wants to look at Venus during its fullest phase. Explain why that is impossible.

______________________________________________

______________________________________________

______________________________________________

______________________________________________

**2.** Imagine a planet that orbits the sun in 100 days and has one moon. The moon goes through a complete set of phases in 20 days. On this planet, how many months would one year have?

______________________________________________

______________________________________________

______________________________________________

______________________________________________

**3.** Using any of the new terms listed on page 634 of your textbook, construct a **concept map** starting with the term "Time." See how many terms you can include.